THE FUTURE OF HUMANITY

DR. MICHIO KAKU

PROFESSOR OF THEORETICAL PHYSICS
CITY UNIVERSITY OF NEW YORK

THE FUTURE OF HUMANITY

TERRAFORMING MARS, INTERSTELLAR
TRAVEL, IMMORTALITY, AND OUR
DESTINY BEYOND EARTH

RANDOM HOUSE LARGE PRINT

Cover design by Michael J. Windsor
Cover images: galaxy © sripfoto / Shutterstock,
spaceship © DM7 / Shutterstock, planet surface
© Kjpargeter / Shutterstock, UFO © Coneyl Jay /
Stone / Getty Images

The Library of Congress has established a
Cataloging-in-Publication record for this title.

ISBN: 978-0-5255-8953-2

www.penguinrandomhouse.com/
large-print-format-books

FIRST LARGE PRINT EDITION

Printed in the United States of America

10 9 8 7 6 5 4 3 2 1

This Large Print edition published
in accord with the standards of the N.A.V.H.

To my loving wife Shizue,

and my daughters Michelle and Alyson

CONTENTS

ACKNOWLEDGMENTS

I would like to thank the following scientists and experts who have generously given their time and expertise to be interviewed for this book and for my national radio and TV programs. Their knowledge and keen insights into science have helped to make this book possible.

I would also like to thank my agent, Stuart Krichevsky, who has, over all these years, helped to make my books a success. I owe him a great debt of gratitude for all his tireless work. He is always the first person I turn to for sound advice.

I would also like to thank Edward Kasten-meier, my editor at Penguin Random House, for his guidance and comments, which have helped to keep the book focused. As always, his advice has considerably improved the manuscript. His sure hand in editing this book is apparent throughout.

I would like to thank the following pioneers and trailblazers:

Peter Doherty, Nobel laureate, St. Jude Children's Research Hospital

Gerald Edelman, Nobel laureate, Scripps Research Institute

Murray Gell-Mann, Nobel laureate, Santa Fe Institute and Caltech

Walter Gilbert, Nobel laureate, Harvard University

David Gross, Nobel laureate, Kavli Institute for Theoretical Physics

Henry Kendall, Nobel laureate, MIT

Leon Lederman, Nobel laureate, Illinois Institute of Technology

Yoichiro Nambu, Nobel laureate, University of Chicago

Henry Pollack, Intergovernmental Panel on Climate Change, Nobel Peace Prize

Joseph Rotblat, Nobel laureate, St. Bartholomew's Hospital

Steven Weinberg, Nobel laureate, University of Texas at Austin

Frank Wilczek, Nobel laureate, MIT

Amir Aczel, author of **Uranium Wars**

Buzz Aldrin, astronaut, NASA, second man to walk on the moon

Geoff Andersen, U.S. Air Force Academy, author of **The Telescope**

David Archer, geophysical scientist, University of Chicago, author of **The Long Thaw**

Jay Barbree, coauthor of **Moon Shot**

John Barrow, physicist, Cambridge University, author of **Impossibility**

Marcia Bartusiak, author of **Einstein's Unfinished Symphony**

Jim Bell, astronomer, Cornell University

Gregory Benford, physicist, University of California, Irvine

James Benford, physicist, president of Microwave Sciences

Jeffrey Bennett, author of **Beyond UFOs**

Bob Berman, astronomer, author of **Secrets of the Night Sky**

Leslie Biesecker, senior investigator, medical genomics, National Institutes of Health

Piers Bizony, author of **How to Build Your Own Spaceship**

Michael Blaese, senior investigator, National Institutes of Health

Alex Boese, founder of Museum of Hoaxes

Nick Bostrom, transhumanist, Oxford
University

Lt. Col. Robert Bowman, director, Institute
for Space and Security Studies

Travis Bradford, author of **Solar Revolution**

Cynthia Breazeal, codirector, Center
for Future Storytelling, MIT Media
Laboratory

Lawrence Brody, senior investigator, medical
genomics, National Institutes of Health

Rodney Brooks, former director, MIT
Artificial Intelligence Laboratory

Lester Brown, founder and president of
Earth Policy Institute

Michael Brown, astronomer, Caltech

James Canton, author of **The Extreme
Future**

Arthur Caplan, founder of Division of
Medical Ethics, NYU School of Medicine

Fritjof Capra, author of **The Science of
Leonardo**

Sean Carroll, cosmologist, Caltech

Andrew Chaikin, author of **A Man on
the Moon**

Leroy Chiao, astronaut, NASA

Eric Chivian, physician, International
Physicians for the Prevention of
Nuclear War

Deepak Chopra, author of **Super Brain**

George Church, professor of genetics, Harvard Medical School

Thomas Cochran, physicist, Natural Resources Defense Council

Christopher Cokinos, astronomer, author of **The Fallen Sky**

Francis Collins, director, National Institutes of Health

Vicki Colvin, chemist, Rice University

Neil Comins, physicist, University of Maine, author of **The Hazards of Space Travel**

Steve Cook, Marshall Space Flight Center, NASA spokesperson

Christine Cosgrove, coauthor of **Normal at Any Cost**

Steve Cousins, Willow Garage Personal Robots Program

Philip Coyle, former U.S. assistant secretary of defense

Daniel Crevier, computer scientist, CEO of Coreco Imaging

Ken Croswell, astronomer, author of **Magnificent Universe**

Steven Cummer, computer scientist, Duke University

Mark Cutkosky, mechanical engineer, Stanford University

Paul Davies, physicist, author of **Superforce**

Daniel Dennett, codirector, Center for
Cognitive Studies, Tufts University

Michael Dertouzos, computer scientist, MIT

Jared Diamond, Pulitzer Prize
winner, UCLA

Mariette DiChristina, editor in chief,
Scientific American

Peter Dilworth, research scientist, MIT
Artificial Intelligence Laboratory

John Donoghue, creator of BrainGate,
Brown University

Ann Druyan, writer and producer, Cosmos
Studios

Freeman Dyson, physicist, Institute for
Advanced Study, Princeton

David Eagleman, neuroscientist, Stanford
University

Paul Ehrlich, environmentalist, Stanford
University

John Ellis, physicist, CERN

Daniel Fairbanks, geneticist, Utah Valley
University, author of **Relics of Eden**

Timothy Ferris, writer and producer, author
of **Coming of Age in the Milky Way**

Maria Finitzo, filmmaker, stem cell expert,
Peabody Award winner

Robert Finkelstein, robotics and computer
science, Robotic Technology, Inc.

Christopher Flavin, senior fellow,
Worldwatch Institute

Louis Friedman, cofounder, Planetary
Society

Jack Gallant, neuroscientist, University of
California, Berkeley

James Garvin, chief scientist, NASA

Evalyn Gates, Cleveland Museum of Natural
History, author of **Einstein's Telescope**

Michael Gazzaniga, neurologist, University
of California, Santa Barbara

Jack Geiger, cofounder, Physicians for Social
Responsibility

David Gelernter, computer scientist, Yale
University

Neil Gershenfeld, director, Center for Bits
and Atoms, MIT Media Laboratory

Paul Gilster, author of **Centauri Dreams**

Rebecca Goldburg, environmentalist, Pew
Charitable Trusts

Don Goldsmith, astronomer, author of **The
Runaway Universe**

David Goodstein, former vice provost,
Caltech

J. Richard Gott III, physicist, Princeton
University, author of **Time Travel in
Einstein's Universe**

Stephen Jay Gould, biologist, Harvard
University

Ambassador Thomas Graham, arms control
and nonproliferation expert under six
presidents

John Grant, author of **Corrupted Science**

Eric Green, director, National Human Genome Research Institute, National Institutes of Health

Ronald Green, genomics and bioethics, Dartmouth College, author of **Babies by Design**

Brian Greene, physicist, Columbia University, author of **The Elegant Universe**

Alan Guth, physicist, MIT, author of **The Inflationary Universe**

William Hanson, author of **The Edge of Medicine**

Chris Hadfield, astronaut, CSA

Leonard Hayflick, University of California, San Francisco School of Medicine

Donald Hillebrand, director, Argonne National Laboratory's Energy Systems Division

Allan Hobson, psychiatrist, Harvard University

Jeffrey Hoffman, astronaut, NASA, MIT

Douglas Hofstadter, Pulitzer Prize winner, author of **Gödel, Escher, Bach**

John Horgan, journalist, Stevens Institute of Technology, author of **The End of Science**

Jamie Hyneman, host of **MythBusters**

Chris Impey, astronomer, University of Arizona, author of **The Living Cosmos**

Robert Irie, computer scientist, the Cog
  Project, MIT Artificial Intelligence
  Laboratory
P. J. Jacobowitz, journalist, **PC Magazine**
Jay Jaroslav, Human Intelligence Enterprise,
  MIT Artificial Intelligence Laboratory
Donald Johanson, paleoanthropologist,
  Institute of Human Origins, discoverer
  of Lucy
George Johnson, science journalist, **New
  York Times**
Tom Jones, astronaut, NASA
Steve Kates, astronomer, TV host
Jack Kessler, professor of medicine,
  Northwestern Medical Group
Robert Kirshner, astronomer, Harvard
  University
Kris Koenig, astronomer, filmmaker
Lawrence Krauss, physicist, Arizona State
  University, author of **The Physics of
  Star Trek**
Lawrence Kuhn, filmmaker, **Closer to Truth**
Ray Kurzweil, inventor and futurist, author
  of **The Age of Spiritual Machines**
Geoffrey Landis, physicist, NASA
Robert Lanza, biotechnology expert, head of
  Astellas Global Regenerative Medicine
Roger Launius, coauthor of **Robots in Space**
Stan Lee, creator of Marvel Comics and
  Spider-Man

Michael Lemonick, former senior science editor, **Time** magazine

Arthur Lerner-Lam, geologist and volcanist, Earth Institute

Simon LeVay, author of **When Science Goes Wrong**

John Lewis, astronomer, University of Arizona

Alan Lightman, physicist, MIT, author of **Einstein's Dreams**

Dan Linehan, author of **SpaceShipOne**

Seth Lloyd, mechanical engineer and physicist, MIT, author of **Programming the Universe**

Werner R. Loewenstein, former director of Cell Physics Laboratory, Columbia University

Joseph Lykken, physicist, Fermi National Accelerator Laboratory

Pattie Maes, professor of media arts and sciences, MIT Media Laboratory

Robert Mann, author of **Forensic Detective**

Michael Paul Mason, author of **Head Cases**

W. Patrick McCray, author of **Keep Watching the Skies!**

Glenn McGee, author of **The Perfect Baby**

James McLurkin, computer scientist, Rice University

Paul McMillan, director, Space Watch

Fulvio Melia, astrophysicist, University of
Arizona

William Meller, author of **Evolution Rx**

Paul Meltzer, Center for Cancer Research,
National Institutes of Health

Marvin Minsky, computer scientist, MIT,
author of **The Society of Mind**

Hans Moravec, Robotics Institute of Carnegie
Mellon University, author of **Robot**

Philip Morrison, physicist, MIT

Richard Muller, astrophysicist, University of
California, Berkeley

David Nahamoo, IBM Fellow, IBM Human
Language Technologies Group

Christina Neal, volcanist, U.S. Geological
Survey

Michael Neufeld, author of **Von Braun:
Dreamer of Space, Engineer of War**

Miguel Nicolelis, neuroscientist, Duke
University

Shinji Nishimoto, neurologist, University of
California, Berkeley

Michael Novacek, paleontology, American
Museum of Natural History

S. Jay Olshansky, biogerontology, University
of Illinois at Chicago, coauthor of **The
Quest for Immortality**

Michael Oppenheimer, environmentalist,
Princeton University

Dean Ornish, clinical professor of medicine, University of California at San Francisco

Peter Palese, virologist, Icahn School of Medicine at Mount Sinai

Charles Pellerin, former director of astrophysics, NASA

Sidney Perkowitz, author of **Hollywood Science**

John Pike, director, GlobalSecurity.org

Jena Pincott, author of **Do Gentlemen Really Prefer Blondes?**

Steven Pinker, psychologist, Harvard University

Tomaso Poggio, cognitive scientist, MIT

Corey Powell, editor in chief, **Discover**

John Powell, founder, JP Aerospace

Richard Preston, author of **The Hot Zone** and **The Demon in the Freezer**

Raman Prinja, astronomer, University College London

David Quammen, evolutionary biologist, author of **The Reluctant Mr. Darwin**

Katherine Ramsland, forensic scientist, DeSales University

Lisa Randall, physicist, Harvard University, author of **Warped Passages**

Sir Martin Rees, astronomer, Cambridge University, author of **Before the Beginning**

Jeremy Rifkin, founder, Foundation on Economic Trends

David Riquier, writing instructor/teaching assistant, Harvard University

Jane Rissler, former senior scientist, Union of Concerned Scientists

Joseph Romm, Senior Fellow at Center for American Progress, author of **Hell and High Water**

Steven Rosenberg, head of Tumor Immunology Section, National Institutes of Health

Oliver Sacks, neurologist, Columbia University

Paul Saffo, futurist, Stanford University and Institute for the Future

Carl Sagan, astronomer, Cornell University, author of **Cosmos**

Nick Sagan, coauthor of **You Call This the Future?**

Michael H. Salamon, NASA's Discipline Scientist for Fundamental Physics and Beyond Einstein program

Adam Savage, host of **MythBusters**

Peter Schwartz, futurist, founder of Global Business Network

Sara Seager, astronomer, MIT

Charles Seife, author of **Sun in a Bottle**

Michael Shermer, founder, Skeptic Society and **Skeptic** magazine

Donna Shirley, former manager, NASA Mars Exploration Program

Seth Shostak, astronomer, SETI Institute

Neil Shubin, evolutionary biologist, University of Chicago, author of **Your Inner Fish**

Paul Shuch, aerospace engineer, executive director emeritus, SETI League

Peter Singer, author of **Wired for War**

Simon Singh, writer and producer, author of **Big Bang**

Gary Small, coauthor of **iBrain**

Paul Spudis, geologist and lunar scientist, author of **The Value of the Moon**

Steven Squyres, astronomer, Cornell University

Paul Steinhardt, physicist, Princeton University, coauthor of **Endless Universe**

Jack Stern, stem cell surgeon, clinical professor of neurosurgery, Yale University

Gregory Stock, UCLA, author of **Redesigning Humans**

Richard Stone, science journalist, **Discover Magazine**

Brian Sullivan, astronomer, Hayden Planetarium

Michael Summers, astronomer, coauthor of **Exoplanets**

Leonard Susskind, physicist, Stanford University

Daniel Tammet, author of **Born on a Blue Day**

Geoffrey Taylor, physicist, University of
Melbourne
Ted Taylor, physicist, designer of U.S.
nuclear warheads
Max Tegmark, cosmologist, MIT
Alvin Toffler, futurist, author of **The
Third Wave**
Patrick Tucker, futurist, World Future Society
Chris Turney, climatologist, University of
Wollongong, author of **Ice, Mud and
Blood**
Neil deGrasse Tyson, astronomer, director,
Hayden Planetarium
Sesh Velamoor, futurist, Foundation for the
Future
Frank von Hippel, physicist, Princeton
University
Robert Wallace, coauthor of **Spycraft**
Peter Ward, coauthor of **Rare Earth**
Kevin Warwick, human cyborg expert,
University of Reading
Fred Watson, astronomer, author of
Stargazer
Mark Weiser, research scientist, Xerox PARC
Alan Weisman, author of **The World
Without Us**
Spencer Wells, geneticist and producer,
author of **The Journey of Man**
Daniel Werthheimer, astronomer, SETI@
home, University of California, Berkeley

Mike Wessler, Cog Project, MIT Artificial
  Intelligence Laboratory
Michael West, CEO, AgeX Therapeutics
Roger Wiens, astronomer, Los Alamos
  National Laboratory
Arthur Wiggins, physicist, author of **The Joy
  of Physics**
Anthony Wynshaw-Boris, geneticist, Case
  Western Reserve University
Carl Zimmer, biologist, coauthor of
  Evolution
Robert Zimmerman, author of **Leaving
  Earth**
Robert Zubrin, founder, Mars Society

THE FUTURE OF HUMANITY

PROLOGUE

One day about seventy-five thousand years ago, humanity almost died.

A titanic explosion in Indonesia sent up a colossal blanket of ash, smoke, and debris that covered thousands of miles. The eruption of Toba was so violent that it ranks as the most powerful volcanic event in the last twenty-five million years. It blew an unimaginable 670 cubic miles of dirt into the air. This caused large areas of Malaysia and India to be smothered by volcanic ash up to thirty feet thick. The toxic smoke and dust eventually sailed over Africa, leaving a trail of death and destruction in its wake.

Imagine, for a moment, the chaos caused by this cataclysmic event. Our ancestors were terrorized by the searing heat and the clouds of gray ash that darkened the sun. Many were choked and poisoned by the thick soot and dust. Then, temperatures plunged, causing a "volcanic winter." Vegetation and wildlife died off as far as the eye could see, leaving only a bleak, desolate landscape. People and animals were left to scavenge the devastated terrain for tiny scraps of food, and most humans died of starvation. It looked as if the entire Earth was dying. The few who survived had only one goal: to flee as far as they could from the curtain of death that descended on their world.

Stark evidence of this cataclysm may perhaps be found in our blood.

Geneticists have noticed the curious fact that any two humans have almost identical DNA. By contrast, any two chimpanzees can have more genetic variation between them than is found in the entire human population. Mathematically, one theory to explain this phenomenon is to assume that, at the time of the explosion, most humans were wiped out, leaving only a handful of us—about two thousand people. Remarkably, this dirty, raggedy band of humans would become the ancestral Adams and Eves who would eventually populate the entire planet. All of us are almost clones of one another, brothers and sisters descended from a tiny, hardy group of humans who could have easily fit inside a modern hotel ballroom.

As they trekked across the barren landscape, they had no idea that one day, their descendants would dominate every corner of our planet.

Today, as we gaze into the future, we see that the events that took place seventy-five thousand years ago may actually be a dress rehearsal for future catastrophes. I was reminded of this in 1992, when I heard the astounding news that, for the first time, a planet orbiting a distant star had been found. With this discovery, astronomers could prove that planets existed beyond our solar system. This was a major paradigm shift in our understanding of the universe. But I was saddened when I heard the next piece of news: this alien planet was orbiting a dead star, a pulsar, that had exploded in a supernova, probably killing everything that might have lived on that planet. No living thing known to science can withstand the withering blast of nuclear energy that emerges when a star explodes close by.

I then imagined a civilization on that planet, aware that their mother sun was dying, working urgently to assemble a huge armada of spaceships that might transport them to another star system. There would have been utter chaos on the planet as people, in panic and desperation, tried to scramble and secure the last few seats on the departing vessels. I imagined the horror felt by those who were left behind to meet their fate as their sun exploded.

It is as inescapable as the laws of physics that humanity will one day confront some type of extinction-level event. But will we, like our ancestors, have the drive and determination to survive and even flourish?

If we scan all the life-forms that have ever existed on the Earth, from microscopic bacteria to towering forests, lumbering dinosaurs, and enterprising humans, we find that more than 99.9 percent of them eventually became extinct. This means that extinction is the norm, that the odds are already stacked heavily against us. When we dig beneath our feet into the soil to unearth the fossil record, we see evidence of many ancient life-forms. Yet only the smallest handful survive today. Millions of species have appeared before us; they had their day in the sun, and then they withered and died. That is the story of life.

No matter how much we may treasure the sight of dramatic, romantic sunsets, the smell of fresh ocean breezes, and the warmth of a summer's day, one day it will all end, and the planet will become inhospitable to human life. Nature will eventually turn on us, as it did to all those extinct life-forms.

The grand history of life on Earth shows that, faced with a hostile environment, organisms inevitably meet one of three fates. They can leave that environment, they can adapt to it, or they will die. But if we look far enough into the future, we will eventually face a disaster so great that adaptation will be virtually impossible. Either we must leave the Earth or we will perish. There is no other way.

These disasters have happened repeatedly in the past, and they will inevitably happen in the future. The Earth has already sustained five major extinction cycles, in which up to 90 percent of all life-forms vanished from the Earth. As sure as day follows night, there will be more to come.

On a scale of decades, we face threats that are not natural but are largely self-inflicted, due to our own folly and shortsightedness. We face the danger of global warming, when the atmosphere of the Earth itself turns against us. We face the danger of modern warfare, as nuclear weapons proliferate in some of the most unstable regions of the globe. We face the danger of weaponized microbes, such as airborne AIDS or Ebola, which can be transmitted by a simple cough or sneeze. This could wipe out upward of 98 percent of the human race. Furthermore, we face an expanding population that consumes resources at a furious rate. We may exceed the carrying capacity of Earth at some point and find ourselves in an ecological Armageddon, vying for the planet's last remaining supplies.

In addition to calamities that we create ourselves, there are also natural disasters over which we have little control. On a scale of thousands of years, we face the onset of another ice age. For the past one hundred thousand years, much of Earth's surface was blanketed by up to a half mile of solid ice. The bleak frozen landscape drove many animals to extinction. Then, ten thousand years ago, there was a thaw in the weather. This brief warming spell led to the sudden rise of modern civilization, and humans have taken advantage of it to spread and thrive. But this flowering has occurred during an interglacial period, meaning we will likely meet another ice age within the next ten thousand years. When it comes, our cities will disappear under mountains of snow and civilization will be crushed under the ice.

We also face the possibility that the supervolcano under Yellowstone National Park may awaken from its long slumber, tearing the United States apart and engulfing the Earth in a choking, poisonous cloud of soot and debris. Previous eruptions took place 630,000, 1.3 million, and 2.1 million years ago. Each event was separated by roughly 700,000 years; therefore, we may be due for another colossal eruption in the next 100,000 years.

On a scale of millions of years, we face the threat of another meteor or cometary impact, similar to the one that helped to destroy the dinosaurs 65 million years ago. Back then, a rock about six miles across plunged into the Yucatán peninsula of Mexico, sending into the sky fiery debris that rained back on Earth. As with the explosion at Toba, only much larger, the ash clouds eventually darkened the sun and led temperatures to plunge globally. With the withering of vegetation, the food chain collapsed. Plant-eating dinosaurs starved to death, followed soon by their carnivorous cousins. In the end, 90 percent of all life-forms on Earth perished in the wake of this catastrophic event.

For millennia, we have been blissfully ignorant of the reality that the Earth is floating in a swarm of potentially deadly rocks. Only within the last decade have scientists begun to quantify the real risk of a major impact. We now know that there are several thousand NEOs (near-Earth objects) that cross the orbit of the Earth and pose a danger to life on our planet. As of June 2017, 16,294 of these objects have been catalogued. But these are just the ones we've found. Astronomers estimate that there are perhaps several million uncharted objects in the solar system that pass by the Earth.

I once interviewed the late astronomer Carl

Sagan about this threat. He stressed to me that "we live in a cosmic shooting gallery," surrounded by potential hazards. It is only a matter of time, he told me, before a large asteroid hits the Earth. If we could somehow illuminate these asteroids, we would see the night sky filled with thousands of menacing points of light.

Even assuming we avoid all these dangers, there is another that dwarfs all the others. Five billion years from now, the sun will expand into a giant red star that fills the entire sky. The sun will be so gigantic that the orbit of the Earth will be inside its blazing atmosphere, and the blistering heat will make life impossible within this inferno.

Unlike all other life-forms on this planet, which must passively await their fate, we humans are masters of our own destiny. Fortunately, we are now creating the tools that will defy the odds given to us by nature, so that we don't become one of the 99.9 percent of life-forms destined for extinction. In this book, we will encounter the pioneers who have the energy, the vision, and the resources to change the fate of humanity. We will meet the dreamers who believe that humanity can live and thrive in outer space. We will analyze the revolutionary advances in technology that will make it possible to leave the Earth and to settle elsewhere in the solar system, and even beyond.

But if there is one lesson we can learn from our history, it is that humanity, when faced with life-threatening crises, has risen to the challenge and has reached for even higher goals. In some sense, the spirit of exploration is in our genes and hardwired into our soul.

But now we face perhaps the greatest challenge of all: to leave the confines of the Earth and soar into outer space. The laws of physics are clear; sooner or later we will face global crises that threaten our very existence.

Life is too precious to be placed on a single planet, to be at the mercy of these planetary threats.

We need an insurance policy, Sagan told me. He concluded that we should become a "two planet species." In other words, we need a backup plan.

In this book, we will explore the history, the challenges, and the possible solutions that lie before us. The path will not be easy, and there will be setbacks, but we have no choice.

From near extinction approximately seventy-five thousand years ago, our ancestors ventured forth and began the colonization of the entire Earth. This book will, I hope, lay out the steps necessary to conquer these obstacles that we will inevitably face in the future. Perhaps our fate is to become a multiplanet species that lives among the stars.

If our long-term survival is at stake, we have a basic responsibility to our species to venture to other worlds.
—CARL SAGAN

The dinosaurs became extinct because they didn't have a space program. And if we become extinct because we don't have a space program, it'll serve us right.
—LARRY NIVEN

INTRODUCTION TOWARD A MULTIPLANET SPECIES

When I was a child, I read Isaac Asimov's Foundation Trilogy, which is celebrated as one of the greatest sagas in the history of science fiction. I was stunned that Asimov, instead of writing about ray gun battles and space wars with aliens, asked a simple but profound question: Where will human civilization be fifty thousand years into the future? What is our ultimate destiny?

In his groundbreaking trilogy, Asimov painted a picture of humanity spread out across the Milky Way, with millions of inhabited planets held together by a vast Galactic Empire. We had traveled so far that the location of the original homeland that gave birth to this great civilization was lost in the mists of prehistory. And there were so many advanced societies distributed throughout the galaxy, with so many people bound together through a complex web of economic ties, that, with this huge sample size, it was possible to use mathematics to predict the future course of events, as if predicting the motion of molecules.

Years ago, I invited Dr. Asimov to speak at our university. Listening to his thoughtful words, I was surprised at his breadth of knowledge. I then asked him a question that had intrigued me since childhood: What had inspired him to write the Foundation series? How had he come up with a theme so large that it embraced the entire galaxy? Without hesitation, he responded that he was inspired by the rise and fall of the Roman Empire. In the story of the empire, one could see how the destiny of the Roman people played out over its turbulent history.

I began to wonder whether the history of humanity has a destiny as well. Perhaps our fate is to eventually create a civilization that spans the entire Milky Way galaxy. Perhaps our destiny is truly in the stars.

Many of the themes underlying Asimov's work were explored even earlier, in Olaf Stapledon's seminal novel **Star Maker.** In the novel, our hero daydreams that he somehow soars into outer space until he reaches faraway planets. Racing across the galaxy as pure consciousness, wandering from star system to star system, he witnesses fantastic alien empires. Some of them rise to greatness, ushering in an era of peace and plenty, and some even create interstellar empires with their starships. Others fall into ruin, wracked by bitterness, strife, and war.

Many of the revolutionary concepts in Stapledon's novel were incorporated into subsequent science fiction. For example, our hero in **Star Maker** discovers that many superadvanced civilizations deliberately keep their existence a secret from lower civilizations, to prevent accidentally contaminating them with advanced technology. This concept is similar to the Prime Directive, one of the guiding principles of the Federation in the **Star Trek** series.

Our hero also comes across a civilization so sophisticated that its members enclose their mother sun in a gigantic sphere to utilize all its energy.

This concept, which would later be called the Dyson sphere, is now a staple of science fiction.

He meets a race of individuals who are in constant telepathic contact with one another. Every individual knows the intimate thoughts of the others. This idea predates the Borg of **Star Trek,** where individuals are connected mentally and are subordinate to the will of the Hive.

And at the end of the novel, he encounters the Star Maker himself, a celestial being who creates and tinkers with entire universes, each with its own laws of physics. Our universe is just one in a multiverse. In total awe, our hero witnesses the Star Maker at work as he conjures up new and exciting realms, discarding those not pleasing to him.

Stapledon's trailblazing novel came as quite a shock in a world where the radio was still considered a miracle of technology. In the 1930s, the idea of achieving a space-faring civilization seemed preposterous. Back then, propeller-driven airplanes were state-of-the-art and had hardly managed to venture above the clouds, so the possibility of traveling to the stars seemed hopelessly remote.

Star Maker was an instant success. Arthur C. Clarke called it one of the finest works of science fiction ever published. It fired up the imagination of a whole generation of postwar science fiction writers. But among the general public, the novel was soon forgotten amidst the chaos and savagery of World War II.

FINDING NEW PLANETS IN SPACE

Now that the Kepler spacecraft and teams of Earth-bound astronomers have discovered about four thousand planets orbiting other stars in the Milky Way galaxy, one begins to wonder if the civilizations described by Stapledon actually exist.

In 2017, NASA scientists identified not one but seven Earth-sized planets orbiting a nearby star, a mere thirty-nine light-years from Earth. Of these seven planets, three of them are close enough to their mother star to support liquid water. Very soon, astronomers will be able to confirm whether or not these and other planets have atmospheres containing water vapor. Since water is the "universal solvent" capable of being the mixing bowl for the organic chemicals that make up the DNA molecule, scientists may be able to show that the conditions for life are common in the universe. We may be on the verge of finding the Holy Grail of planetary astronomy, a twin of the Earth in outer space.

Around the same time, astronomers made another game-changing discovery, an Earth-sized planet named Proxima Centauri b, which orbits the star closest to our sun, Proxima Centauri, which is just 4.2 light-years away from us. Scientists have long conjectured that this star would be one of the first to be explored.

These planets are just a few of the recent entries in the huge Extrasolar Planets Encyclopaedia, which has to be updated practically every week. It contains strange, unusual star systems that Stapledon could only have dreamt of—including systems where four or more stars rotate among one another. Many astronomers believe that if you can imagine any bizarre formation of planets, then it probably exists somewhere in the galaxy, as long as it doesn't violate some law of physics.

This means that we can roughly calculate how many Earth-sized planets there are in the galaxy. Since it has about one hundred billion stars, there might be twenty billion Earth-sized planets orbiting a sun-like star in our galaxy alone. And since there are one hundred billion galaxies that can be seen with our instruments, we can estimate how many Earth-sized planets there are in the visible universe: a staggering two billion trillion.

Realizing that the galaxy could be teeming with habitable planets, you will never see the night sky in the same way again.

Once astronomers have identified these Earth-sized planets, the next goal will be to analyze their atmospheres for oxygen and water vapor, a sign of life, and listen for radio waves, which would signal the existence of an intelligent civilization. Such a discovery would be one of the great turning points in human history, comparable to the taming of fire. Not only would it redefine our relationship to the rest of the universe, it would also change our destiny.

THE NEW GOLDEN AGE OF SPACE EXPLORATION

All these exciting discoveries of exoplanets, along with the novel ideas brought about by a fresh new generation of visionaries, are rekindling the public's interest in space travel. Originally, what drove the space program was the Cold War and superpower rivalry. The public did not mind spending a staggering 5.5 percent of the nation's federal budget on the Apollo space program because our national prestige was at stake. However, this feverish competition could not be sustained forever, and the funding eventually collapsed.

U.S. astronauts last walked on the surface of the moon about forty-five years ago. Now, the Saturn V rocket and the space shuttle are dismantled and rusting in pieces in museums and junkyards, their stories languishing in dusty history books. In the years that followed, NASA was criticized as the "agency to nowhere." It has been spinning its wheels for decades, boldly going where everyone has gone before.

But the economic situation has begun to change. The price of space travel, once so high it could cripple a nation's budget, has been dropping steadily, in large part because of the influx of energy, money, and enthusiasm from a rising cohort of entrepreneurs. Impatient with NASA's sometimes glacial pace, billionaires like Elon Musk, Richard Branson, and Jeff Bezos have been opening up their checkbooks to build new rockets. Not only do they want to turn a profit, they also want to fulfill their childhood dreams of going to the stars.

Now there is a rejuvenated national will. The question is no longer whether the U.S. will send astronauts to the Red Planet, but when. Former president Barack Obama stated that astronauts would walk on the surface of Mars sometime after 2030, and President Donald Trump has asked NASA to accelerate that timetable.

A fleet of rockets and space modules capable of an interplanetary journey—like NASA's Space Launch System (SLS) booster rocket with the Orion capsule and Elon Musk's Falcon Heavy booster rocket with the Dragon capsule—are in the early testing phase. They will do the heavy lifting, taking our astronauts to the moon, asteroids, Mars, and even beyond. In fact, so much publicity and enthusiasm have been generated by this mission that there is rivalry building up around it. Perhaps there will be a traffic jam over Mars as different groups compete to plant the first flag on Martian soil.

Some have written that we are entering a new golden age of space travel, when exploring the universe will once again become an exciting part of the national agenda after decades of neglect.

As we look to the future, we can see the outlines of how science will transform space exploration. Because of revolutionary advances in a wide range of modern technologies, we can now speculate how our civilization may one day move into outer space, terraforming planets and traveling among the stars. Although this is a long-term goal, it is now possible to give a reasonable time frame and estimate when certain cosmic milestones will be met.

In this book, I will investigate the steps necessary to accomplish this ambitious goal. But the key to discovering how our future may unfold is to understand the science behind all of these miraculous developments.

REVOLUTIONARY WAVES OF TECHNOLOGY

Given the vast frontiers of science that lie before us, it may help to put the broad panorama of human history into perspective. If our ancestors could see us today, what would they think? For most of human history, we lived wretched lives, struggling in a hostile, uncaring world where life expectancy was between twenty and thirty years of age. We were mostly nomads, carrying all our possessions on our backs. Every day was a struggle to secure food and shelter. We lived in constant fear of vicious predators, disease, and hunger. But if our ancestors could see us today, with our ability to send images instantly across the planet, with rockets that can take us to the moon and beyond, and with cars that can drive themselves, they would consider us to be sorcerers and magicians.

History reveals that scientific revolutions come in waves, often stimulated by advances in physics. In the nineteenth century, the first wave of science and technology was made possible by physicists who created the theory of mechanics and thermodynamics. This enabled engineers to produce the steam engine, leading to the locomotive and the industrial revolution. This profound shift in technology lifted civilization from the curse of ignorance, backbreaking labor, and poverty and took us into the machine age.

In the twentieth century, the second wave was spearheaded by physicists who mastered the laws of electricity and magnetism, which in turn ushered in the electric age. This made possible the electrification of our cities with the advent of dynamos, generators, TV, radio, and radar. The second wave gave birth to the modern space program, which took us to the moon.

In the twenty-first century, the third wave of science has been expressed in high tech, spearheaded by the quantum physicists who invented the transistor and the laser. This made possible the supercomputer, the internet, modern telecommunications, GPS, and the explosion of the tiny chips that have permeated every aspect of our lives.

In this book, I will describe the technologies that will take us even farther as we explore the planets and the stars. In part 1, we will discuss the effort to create a permanent moon base and to colonize and terraform Mars. To do this, we will have to exploit the fourth wave of science, which consists of artificial intelligence, nanotechnology, and biotechnology. The goal of terraforming Mars exceeds our capability today, but the technologies of the twenty-second century will allow us to turn this bleak, frozen desert into a habitable world. We will consider the use of self-replicating robots, superstrong, lightweight nanomaterials, and bioengineered crops to drastically cut costs and make Mars into a veritable paradise. Eventually, we will progress beyond Mars and develop settlements on the asteroids and the moons of the gas giants, Jupiter and Saturn.

In part 2, we will look ahead to a time when we will be able to move beyond the solar system and explore the nearby stars. Again, this mission surpasses our current technology, but fifth wave technologies will make it possible: nanoships, laser sails, ramjet fusion machines, antimatter engines. Already, NASA has funded studies on the physics necessary to make interstellar travel a reality.

In part 3, we analyze what it would require to modify our bodies to enable us to find a new home among the stars. An interstellar journey may take decades or even centuries, so we may have to genetically engineer ourselves to survive for prolonged periods in deep space, perhaps by extending the human life span. Although a fountain of youth is not possible today, scientists are exploring promising avenues that may allow us to slow and perhaps stop the aging process. Our descendants may enjoy some form of immortality. Furthermore, we may have to genetically engineer our bodies to flourish on distant planets with different gravity, atmospheric composition, and ecology.

Thanks to the Human Connectome Project, which will map every neuron in the human brain, one day we may be able to send our connectomes into outer space on giant laser beams, eliminating a number of problems in interstellar travel. I call this laser porting, and it may free our consciousness to explore the galaxy or even the universe at the speed of light, so we don't have to worry about the obvious dangers of interstellar travel.

If our ancestors in the last century would think of us today as magicians and sorcerers, then how might we view our descendants a century from now?

More than likely, we would consider our descendants to be like Greek gods. Like Mercury, they would be able to soar into space to visit nearby planets. Like Venus, they would have perfect immortal bodies. Like Apollo, they would have unlimited access to the sun's energy. Like Zeus, they would be able to issue mental commands and have their wishes come true. And they would be able to conjure up mythical animals like Pegasus using genetic engineering.

In other words, our destiny is to become the gods that we once feared and worshipped. Science will give us the means by which we can shape the universe in our image. The question is whether we will have the wisdom of Solomon to accompany this vast celestial power.

There is also the possibility that we will make contact with extraterrestrial life. We will discuss what might happen were we to encounter a civilization that's a million years more advanced than ours, that has the capability to roam across the galaxy and alter the fabric of space and time. They might be able to play with black holes and use wormholes for faster-than-light travel.

In 2016, speculation about advanced civilizations in space reached a fever pitch among astronomers and the media, with the announcement that astronomers had found evidence of some sort of colossal "megastructure," perhaps as big as a Dyson sphere, orbiting around a distant star many light-years away. While the evidence is far from conclusive, for the first time, scientists were confronted with evidence that an advanced civilization may actually exist in outer space.

Lastly, we explore the possibility that we will face not just the death of the Earth but the death of the universe itself. Although our universe is still young, we can foresee the day in the distant future when we might approach the Big Freeze as temperatures plunge to near absolute zero and all life as we know it will likely cease to exist. At that point, our technology might be advanced enough to leave the universe and venture through hyperspace to a new, younger universe.

Theoretical physics (my own specialization) opens up the notion that our universe could be just a single bubble floating in a multiverse of other bubble universes. Perhaps among the other universes in the multiverse, there is a new home for us. Gazing upon the multitude of universes, perhaps we will be able to reveal the grand designs of a Star Maker.

So the fantastic feats of science fiction, once considered the byproduct of the overheated imagination of dreamers, may one day become reality.

Humanity is about to embark on perhaps its greatest adventure. And the gap that separates the speculations of Asimov and Stapledon from reality may be bridged by the astonishing and rapid advancements being made in science. And the first step we take in our long journey to the stars begins when we leave the Earth. As the old Chinese proverb says, the journey of a thousand miles begins with the first step. The journey to the stars begins with the very first rocket.

PART I LEAVING
THE EARTH

Anyone who sits on top of the largest hydrogen-oxygen fueled system in the world, knowing they're going to light the bottom, and doesn't get a little worried, does not fully understand the situation.
—ASTRONAUT JOHN YOUNG

1 PREPARING FOR LIFTOFF

On October 19, 1899, a seventeen-year-old boy climbed a cherry tree and had an epiphany. He had just read H. G. Wells's **War of the Worlds** and was excited by the idea that rockets could allow us to explore the universe. He imagined how wonderful it would be to make some device that had even the **possibility** of traveling to Mars and had a vision that it was our destiny to explore the Red Planet. By the time he came down from that tree, his life had been forever changed. That boy would dedicate his life to the dream of perfecting a rocket that would make this vision a reality. He would celebrate October 19 for the rest of his life.

His name was Robert Goddard, and he went on to perfect the first liquid fueled multistage rocket, setting into motion events that changed the course of human history.

TSIOLKOVSKY—A LONELY VISIONARY

Goddard was one of a handful of pioneers who, despite isolation, poverty, and ridicule from their peers, forged ahead against all odds and laid the foundation for space travel. One of the first of these visionaries was the great Russian rocket scientist Konstantin Tsiolkovsky, who mapped out the theoretical basis for space travel and paved the way for Goddard. Tsiolkovsky lived in total poverty, was a recluse, and scraped by as a school-teacher. As a youth, he spent most of his time in the library, devouring science journals, learning Newton's laws of motion, and applying them to space travel. His dream was to travel to the moon and Mars. On his own, without the help of the scientific community, he figured out the mathematics, physics, and mechanics of rockets, and he calculated the escape velocity of the Earth—

that is, the speed necessary to escape the gravity of the Earth—to be twenty-five thousand miles per hour, which is far greater than the fifteen miles per hour one could attain with horses in his time.

In 1903, he published his famous rocket equation, which allows one to determine the maximum velocity of a rocket, given its weight and fuel supply. The equation revealed that the relationship between speed and fuel is exponential. Normally, one might assume that if you want to double the velocity of a rocket, you simply need to double the amount of fuel. Instead, the amount of fuel you need rises exponentially with the change in velocity, so that enormous amounts of fuel are needed to give an extra boost in speed.

This exponential relationship made it clear that you would need huge amounts of fuel to leave the Earth. With his formula, Tsiolkovsky was for the first time able to estimate how much fuel was necessary to go to the moon, long before his vision became reality.

Tsiolkovsky's guiding philosophy was, "The Earth is our cradle, but we cannot be in the cradle forever," and he believed in a philosophy called cosmism, which holds that the future of humanity is to explore outer space. In 1911, he wrote, "To place one's feet on the soil of asteroids, to lift a stone from the moon with your hand, to construct moving stations in ether space, to organize inhabited rings around the Earth, Moon and

Sun, to observe Mars at the distance of several tens of miles, to descend to its satellites or even to its own surface—what could be more insane!"

Although Tsiolkovsky was too poor to convert his mathematical equations into actual models, the next step was taken by Robert Goddard, who actually built the prototypes that would one day form the basis of space travel.

ROBERT GODDARD—FATHER OF ROCKETRY

Robert Goddard first became interested in science as a child witnessing the electrification of his hometown. He came to believe that science would revolutionize every aspect of our lives. His father encouraged this interest, buying him a telescope, microscope, and a subscription to **Scientific American.** First he began experimenting with kites and balloons. While reading in the library one day, he stumbled across Isaac Newton's celebrated **Principia Mathematica** and learned the laws of motion. His focus soon became the application of Newton's laws to rocketry.

Goddard systematically turned this curiosity into a usable scientific tool by introducing three innovations. First, Goddard experimented with different types of fuels and realized that powdered fuel is inefficient. The Chinese had invented gunpowder centuries earlier and used it for rockets, but gunpowder burns unevenly and hence rockets remained mainly toys. His first stroke of brilliance was to replace powdered fuel with liquid fuel, which could be precisely controlled so that it burned cleanly and steadily. He built a rocket with two tanks, one containing a fuel, such as alcohol, and the other tank containing an oxidizer, such as liquid oxygen. These liquids were fed by a series of pipes and valves into the firing chamber, creating a carefully controlled explosion that could propel a rocket.

Goddard realized that as the rocket rose into the sky, its fuel tanks were gradually depleted. His next innovation was to introduce multistage rockets that discarded spent fuel tanks and therefore could shed some dead weight along the way, vastly increasing their range and efficiency.

And third, he introduced gyroscopes. Once a gyroscope is sent spinning, its axis always points in the same direction, even if you rotate it. For example, if the axis points toward the North Star, it will continue to point in that direction if you turn it upside down. This means that a spaceship, if it were to wander in its trajectory, can alter its rockets to compensate for this motion and return to its original course. Goddard realized he could use gyroscopes to help keep his rockets on target.

In 1926, he made history with the first successful launch of a liquid fueled rocket. It rose 41 feet into the air, flew for 2.5 seconds, and landed 184 feet away in a cabbage patch. (The exact site is now hallowed ground to every rocket scientist, and it has been declared a National Historic Landmark.)

In his laboratory at Clark College he established the basic architecture for all chemical rockets. The thundering behemoths we see blasting off from launchpads today are direct descendants of the prototypes he built.

FACING RIDICULE

Despite his successes, Goddard proved to be an ideal whipping boy for the media. When word leaked out in 1920 that he was giving serious thought to space travel, the **New York Times** published scathing criticism that would have crushed any lesser scientist. "That Professor Goddard," the **Times** snickered, "with his 'chair' in Clark College . . . does not know the relation of action and reaction, and of the need to have something better than a vacuum against which to react—to say that would be absurd. Of course he only seems to lack the knowledge ladled out daily in high school." And in 1929, after he launched one of his rockets, the local Worcester newspaper ran a degrading headline: "Moon Rocket Misses Target by 238,799 1/2 Miles." Clearly the **Times** and others did not understand Newton's laws of motion and incorrectly believed that rockets could not move in the vacuum of outer space.

Newton's third law, which states that for every action, there is an equal and opposite reaction, governs space travel. This law is known to any child who has ever blown up a balloon, released it, and watched the balloon fly in all directions. The action is the air that suddenly rushes out of the balloon, and the reaction is the forward motion of the balloon itself. Similarly, in a rocket, the action is the hot gas ejected out of one end, while the reaction is the forward motion of the rocket that propels it, even in the vacuum of space.

Goddard died in 1945 and did not live long enough to see the apology written by the editors of the **New York Times** after the Apollo moon landing in 1969. They wrote, "It is now definitely established that a rocket can function in a vacuum as well as in an atmosphere. The **Times** regrets the error."

ROCKETS FOR WAR AND PEACE

In the first phase of rocketry, we had the dreamers, like Tsiolkovsky, who worked out the physics and mathematics of space travel. In the second phase, we had people like Goddard, who actually built the first prototypes of these rockets. In the third phase, rocket scientists caught the eye of major governments. Wernher von Braun would take the sketches, dreams, and models of his predecessors and with the support of the German government—and later the United States— would create gargantuan rockets that would successfully take us to the moon.

The most celebrated of all rocket scientists was born an aristocrat. Baron Wernher von Braun's father was the German minister of agriculture during the Weimar Republic, and his mother could trace her ancestry to the royal houses of France, Denmark, Scotland, and England. Von Braun was an accomplished pianist as a child and even wrote original works of music. At one point, he might have become a renowned musician or composer. But his destiny changed when his mother bought him a telescope. He became fascinated by space. He devoured science fiction and was inspired by the speed records set by rocket-propelled cars. One day, when he was twelve, he unleashed chaos in the crowded streets of Berlin by attaching a series of fireworks to a toy wagon. He was delighted that it took off like, well, a rocket. However, the police were less impressed. Von Braun was taken into custody but released because of his father's influence. As he recalled fondly years later, "It performed beyond my wildest dreams. The wagon careened crazily about, trailing fire like a comet. When the rockets burned out, ending their sparkling performance with a magnificent thunderclap, the wagon rolled majestically to a halt."

Von Braun confessed that he was never good with mathematics. But his drive to perfect rocketry led him to master calculus, Newton's laws, and the mechanics of space travel. As he once told his professor, "I plan on traveling to the Moon."

He became a graduate student in physics and earned his Ph.D. in 1934. But he spent much of his time with the amateur Berlin Rocket Society, an organization that used spare parts to build and test rockets on a deserted three-hundred-acre piece of land outside of the city. That year, the society successfully tested a rocket that rose two miles into the air.

Von Braun might have become a professor of physics at some German university, writing learned articles about astronomy and astronautics. But war was in the air, and all of German society, including the universities, was being militarized. Unlike his predecessor, Robert Goddard, who had requested funding from the U.S. military but was turned down, von Braun got an entirely different reception from the Nazi government. The German Army Ordnance Department, always searching for new weapons of war, noticed von Braun and offered him generous funding. His work was so sensitive that his Ph.D. thesis was classified by the army and wasn't published until 1960.

Von Braun, by all accounts, was apolitical. Rocketry was his passion, and if the government would fund his research, he would accept it. The Nazi Party offered him the dream of a lifetime: directorship of a massive project to build the rocket of the future, with a nearly unlimited budget, employing the cream of German science. Von Braun claimed that being offered membership in the Nazi Party and even the SS was a rite of passage for government workers rather than a reflection of his politics. But when you make a deal with the devil, the devil always asks for more.

RISE OF THE V-2

Under von Braun's leadership, the scribblings and sketches of Tsiolkovsky and the prototypes of Goddard became the Vengeance Weapon 2 rocket, an advanced weapon of war that terrorized London and Antwerp, blowing up entire city blocks. The V-2 was unbelievably powerful. It dwarfed Goddard's rockets, making them look like toys. The V-2 stood forty-six feet tall and weighed 27,600 pounds. It could travel at a blazing speed of 3,580 miles per hour and it achieved a maximum altitude of about sixty miles. It hit its targets at three times the speed of sound, giving no warning apart from a double cracking noise as it broke the sound barrier. And it had an operational range of two hundred miles. Countermeasures were futile since no human could track it and no airplane could catch it.

The V-2 set a number of world records, shattering all past achievements in terms of speed and range for a rocket. It was the first long-range guided ballistic missile. It was the first rocket to break the sound barrier. And most impressively, it was the first rocket ever to leave the boundary of the atmosphere and enter outer space.

The British government was so flummoxed by this advanced weapon that they had no words for it. They invented the story that all these explosions were caused by faulty gas mains. But because the agent of these horrific explosions clearly came from the sky, the public sarcastically referred to them as "flying gas pipes." Only after the Nazis announced that a new weapon of war had been unleashed against the British did Winston Churchill finally admit that England had been attacked by rockets.

Suddenly, it appeared as if the future of Europe, and Western civilization itself, might hinge upon the work of a small, isolated band of scientists led by von Braun.

HORRORS OF WAR

The successes of Germany's advanced weapons came at a tremendous human cost. More than three thousand V-2 rockets were launched against the Allies, resulting in nine thousand deaths. It is estimated that the death toll was even higher—at least twelve thousand—for the prisoners of war who built the V-2 rockets in slave labor camps. The devil wanted its due. Von Braun realized too late that he was in way over his head.

He was horrified when he visited the site where the rockets were built. A friend of von Braun's quoted him as saying, "It is hellish. My spontaneous reaction was to talk to one of the SS guards, only to be told with unmistakable harshness that I should mind my own business, or find myself in the same striped fatigues! . . . I realized that any attempt of reasoning on humane grounds would be utterly futile." Another colleague, when asked if von Braun had ever criticized these death camps, replied, "If he had done it, in my opinion, he would have been shot on the spot."

Von Braun became a pawn of the monster he helped to create. In 1944, when the war effort was in trouble, he got drunk at a party and said that the war was not going well. All he wanted to do was work on rocketry. He regretted that they were working on these weapons of war instead of a spaceship. Unfortunately, there was a spy at the party, and when his drunken comments were relayed to the government, he was arrested by the Gestapo. For two weeks, he was held in a prison cell in Poland, not knowing if he would be shot. Other charges, including rumors that he was a communist sympathizer, were brought to light as Hitler decided his fate. Some officials feared he might defect to England and sabotage the V-2 effort.

Eventually, a direct appeal from Albert Speer to Hitler spared von Braun's life because he was still considered too crucial to the V-2 effort.

The V-2 rocket was decades ahead of its time, but it didn't enter full-time combat until the end of 1944, which was too late to stem the collapse of the Nazi empire, as the Red Army and Allied forces converged on Berlin.

In 1945, von Braun and one hundred of his assistants surrendered to the Allies. They, along with three hundred railroad cars of V-2 rockets and parts, were smuggled back to the U.S. This was part of a program, called Operation Paperclip, to debrief and recruit former Nazi scientists.

The U.S. Army scrutinized the V-2, which eventually became the basis of the Redstone rocket, and von Braun and his assistants had their Nazi records "cleansed." But von Braun's highly ambiguous role in the Nazi government continued to haunt him. The comedian Mort Sahl would summarize his career with the quip, "I reach for the stars, but sometimes I hit London." Singer Tom Lehrer penned the words, "Once the rockets are up, who cares where they come down? That's not my department."

ROCKETRY AND SUPERPOWER RIVALRY

In the 1920s and 1930s, U.S. government officials missed a strategic opportunity when they did not recognize the prophetic work being done in their own backyard by Goddard. They missed a second strategic opportunity after the war, with the arrival of von Braun. In the 1950s, they left von Braun and his assistants in limbo, without giving them any real focus. Eventually, interservice rivalry took over. The army, under von Braun, created the Redstone rocket, while the navy had the Vanguard missile and the air force the Atlas.

Without any immediate obligations for the army, von Braun began to take an interest in science education. He created a series of animated TV specials with Walt Disney that captured the imagination of future rocket scientists. In the series, von Braun painted the broad outlines of a massive scientific effort to land on the moon as well as to develop a fleet of ships that would reach Mars.

While the U.S. rocketry program proceeded by fits and starts, the Russians moved ahead rapidly with theirs. Joseph Stalin and Nikita Khrushchev grasped the strategic importance of the space program and made it a top priority. The Soviet program was put under the direction of Sergei Korolev, whose very identity was kept top secret. For years he was only referred to mysteriously as "Chief Designer" or "the Engineer." The Russians had also captured a number of V-2 engineers and moved them to the Soviet Union. With their guidance, the Soviets took the basic V-2 design and quickly built a series of rockets based on it. Essentially, the entire U.S. and USSR arsenals were based on modifying or lashing together the V-2 rockets, which in turn were based on Goddard's pioneering prototypes.

One of the major goals of both the United States and USSR was launching the first artificial satellite. It was Isaac Newton himself who first proposed the concept. In a now-famous diagram, Newton noted that if you fire a cannonball from a mountaintop, it will fall near the base of the mountain. Following his equations of motion, however, the faster the cannonball travels, the farther it will go. If you fire the cannonball fast enough, it will circle completely around the Earth and become a satellite. Newton made a historic breakthrough: if you replace this cannonball with the moon, then his equations of motion should be able to predict the precise nature of the moon's orbit.

In his cannonball thought experiment, he asked

a key question: If an apple falls, does the moon also fall? Since the cannonball is in free fall as it goes around the Earth, the moon must also be in free fall. Newton's insight set into motion one of the greatest revolutions in all of history. Newton could now calculate the motion of cannonballs, moons, planets—almost everything. For example, using his laws of motion, you can easily show that you must fire the cannonball at eighteen thousand miles per hour in order to have it orbit the Earth.

Newton's vision became a reality when the Soviets launched the world's first artificial satellite, Sputnik, in October 1957.

SPUTNIK AGE

The immense shock to the American psyche upon learning of Sputnik cannot be underestimated. Americans quickly realized that the Soviets led the world in rocket science. The humiliation was made worse when, two months later, the navy's Vanguard missile suffered a catastrophic failure on international TV. I vividly remember, as a child, asking my mother if I could stay up and watch the missile launch. She reluctantly agreed. I was horrified to witness the Vanguard missile rise four feet into the air, then drop back down four feet, tip over, and destroy its own launchpad in a huge, blinding explosion. I could clearly see the nose cone at the top of the missile, which contained the satellite, topple over and disappear in a ball of flames.

The humiliation continued when the second Vanguard launch a few months later also failed. The press had a field day, calling the missile "Flopnik" and "Kaputnik." The Soviet U.N. delegate even joked that Russia should give aid to the United States.

Trying to recover from this huge media blow to our national prestige, von Braun was ordered to quickly launch a satellite, Explorer I, using the Juno I missile. The Juno I was based on the Redstone rocket, which in turn was based on the V-2.

But the Soviets had a series of aces up their sleeve. A sequence of historic "firsts" dominated the headlines for the next several years:

1957: Sputnik 2 carried the first animal, a dog named Laika, into orbit

1957: Lunik 1 was the first rocket to fly past the moon

1959: Lunik 2 was the first to hit the moon

1959: Lunik 3 was the first rocket to photograph the back side of the moon

1960: Sputnik 5 had the first animals returned safely from space

1961: Venera 1 was the first probe to fly past Venus

The Russian space program reached its crowning achievement when Yuri Gagarin safely orbited the Earth in 1961.

I distinctly remember those years, when Sputnik spread panic throughout the United States. How could a seemingly backward nation, the Soviet Union, suddenly leapfrog ahead of us?

Commentators concluded that the root cause of this fiasco was the U.S. education system. American students were falling behind the Soviets. A crash campaign had to be mounted so that money, resources, and media attention could be devoted to producing a new generation of American scientists who could compete with the Russians. Articles at the time declared that "Ivan can read, but Johnny cannot."

Out of this troubled time came the Sputnik generation, a cohort of students who considered it their national duty to become physicists, chemists, or rocket scientists.

Under enormous pressure to let the military wrest control over the U.S. space program from seemingly hapless civilian scientists, President Dwight Eisenhower bravely insisted on continued civilian oversight and created NASA. Then President John F. Kennedy, responding to Gagarin's orbital trip, called for an expedited program to put humans on the moon by the end of the decade.

This call galvanized the nation. By 1966, an astounding 5.5 percent of the U.S. federal budget was going into the lunar program. As always NASA moved cautiously, perfecting the technology needed to bring a moon landing about in a series of launches. First, there was the one-manned craft called Mercury, and then the two-manned Gemini, and finally the three-manned Apollo. NASA also carefully mastered each step in space travel. First, astronauts left the safety of their spaceships and made the first spacewalks. Then astronauts mastered the complex art of docking their spaceship with another ship. Next, astronauts orbited completely around the moon but did not land on the surface. Then, finally, NASA was ready to launch astronauts directly to the moon.

Von Braun was called in to help build the Saturn V, which was to be the biggest rocket ever built. This rocket was a truly marvelous engineering masterpiece. It stood sixty feet taller than the Statue of Liberty. It could lift a payload of 310,000 pounds into orbit around the Earth. Most important, it could send large payloads past twenty-five thousand miles per hour, which is the escape velocity of the Earth.

The possibility of a fatal disaster was ever on the minds of NASA. President Richard Nixon had two speeches prepared for his TV announcement of the results of the Apollo 11 mission. One speech was to report that the effort was a failure and that American astronauts had died on the moon. This scenario actually came very close to happening. In the final seconds before the Lunar Module was to land, computer alarms went off inside the capsule. Neil Armstrong manually took control of the spacecraft and gently landed it on the moon. Analysis later showed that they had only fifty seconds of fuel left; the capsule might have crashed.

Fortunately, on July 20, 1969, President Nixon was able to deliver the other speech, congratulating our astronauts for their successful landing. Even today, the Saturn V is the only rocket ever to carry humans beyond near-Earth orbit. Surprisingly, it performed flawlessly. A total of fifteen Saturn rockets were built, and thirteen were flown, without a mishap. Altogether, the Saturn V sent twenty-four astronauts to either land on or fly by the moon, from December 1968 to December 1972, and the Apollo astronauts were rightly hailed as heroes who had restored our national reputation.

The Russians were also heavily involved in the race to the moon. However, they ran into a number of difficulties. Korolev, who had directed the Soviet rocket program, died in 1966. And there were four failures of the N-1 rocket, which was to take Russian astronauts to the moon. But perhaps most decisive was the fact that the Soviet economy, already stretched by the Cold War, could not compete with the U.S. economy, which was more than twice its size.

LOST IN SPACE

I remember the moment that Neil Armstrong and Buzz Aldrin set foot on the moon. It was July 1969, and I was in the U.S. Army, training with the infantry at Fort Lewis, Washington, and wondering if I would be sent to fight in Vietnam. It was thrilling to know that history was being made right before our eyes, but it was also disconcerting to know that if I died on the battlefield, I would not be able to share my memories of the historic moon landing with my future children.

After the last launch of the Saturn V in 1972, the American public began to be consumed with other matters. The War on Poverty was in full swing, and the Vietnam War was devouring more and more money and lives. Going to the moon seemed like a luxury when Americans were starving next door or dying abroad.

The astronomical cost of the space program was unsustainable. Plans were made for the post-Apollo era. Several proposals were on the table.

One prioritized sending unmanned rockets into space, an effort led by the military, commercial, and scientific groups that were less interested in heroics and more interested in valuable payloads. Another proposal emphasized sending humans into space. The harsh reality was that it was always easier to get Congress and the taxpayer to fund astronauts into space, rather than some nameless space probe. As one congressman summed up, "No Buck Rogers, no bucks."

Both groups wanted quick and cheap access to outer space rather than costly missions that were years apart. But the end result was a strange hybrid that pleased no one. Astronauts would be sent along with freight and cargo.

The compromise took the form of the space shuttle, which began operating in 1981. This craft was an engineering marvel that exploited all the lessons and advanced technologies developed over the past decades. It was capable of sending sixty thousand pounds of payload into orbit and then docking with the International Space Station. Unlike the Apollo space modules, which were retired after each flight, the space shuttle was designed to be partially reusable. It was capable of sending seven astronauts into space and then flying them back home, like an airplane. As a result, space travel gradually started to seem routine. Americans became accustomed to seeing astronauts waving at us from their latest visit to the International Space Station, which itself was a compromise between the many nations paying the bills.

Over time, problems emerged with the space shuttle. For one, although the shuttle was designed to save money, costs nevertheless began to soar, so that each launch consumed about $1 billion. Sending anything into near-Earth orbit on the shuttle cost roughly $40,000 per pound, which was about four times the cost of other delivery systems. Companies complained that it was much cheaper to send their satellites using conventional rockets. Secondly, flights took place infrequently, with many months between launches. Even the U.S. Air Force was frustrated by these limitations and eventually canceled some of its space shuttle launches in favor of using other options.

Physicist Freeman Dyson of the Institute for Advanced Study in Princeton, New Jersey, has his own thoughts on why the space shuttle failed to live up to expectations. When we look at the history of the railroad, we see that it initially started as a carrier for all goods, including humans and commercial products. The commercial side and consumer side of the industry each had their own distinct priorities and concerns, and they eventually split apart, increasing efficiency and lowering costs. The space shuttle, however, never made this split and remained a cross between commercial and consumer interests. Instead of being "everything to everyone," it became "nothing to nobody," especially with its cost overruns and flight delays.

And matters worsened after the **Challenger** and **Columbia** tragedies, which cost the lives of fourteen brave astronauts. These disasters weakened public, private, and government support for the space program. As physicists James and Gregory Benford wrote, "Congress came to see NASA primarily as a jobs program, not an exploratory agency." They also observed that "very little useful science got done in the space station . . . The station was about camping in space, not living in space."

Without the wind of the Cold War in its sails, the space program rapidly lost funding and momentum. Back in the heyday of the Apollo space program, the joke was that NASA could go to Congress asking for funds and say just one word: "Russia!" Then Congress would whip out its checkbook and reply, "How much?" But those days were long gone. As Isaac Asimov said, we scored a touchdown—and then we took our football and went home.

Things finally came to a head in 2011, when former President Barack Obama ordered a new

"Valentine's Day massacre." In one sweeping gesture, he canceled the Constellation program (the replacement for the shuttle), the moon program, and the Mars program. To relieve the tax burden on the public, he defunded these programs in hope that the private sector would make up the difference. Twenty thousand veterans of the space program were suddenly laid off, throwing away the collective wisdom of NASA's best and brightest. The greatest humiliation was that American astronauts, after going toe-to-toe with Russian astronauts for decades, would now be forced to hitchhike on Russian booster rockets. The heyday of space exploration, it seemed, was over; things had reached rock bottom.

The problem could be summed up in one four-letter word, **c-o-s-t**. It takes $10,000 to put a pound of anything in near-Earth orbit. Imagine your body made of solid gold. That's roughly what it would take to put you into orbit. To put something on the moon can easily cost $100,000 per pound. And to put things on Mars costs upward of a million dollars per pound. Estimates of putting an astronaut on Mars are often between $400 and $500 billion in total.

I live in New York City. For me it was a sad day when the space shuttle came to town. Although curious tourists lined up and cheered as the shuttle came rolling down the street, it represented the end of an era. The ship was put on display, eventually resting off the pier on Forty-Second Street. With no replacement in sight, it felt as if we were giving up on science, and hence our future.

Looking back at those dark days, I am sometimes reminded of what happened to the great Chinese imperial fleet in the fifteenth century. Back then, the Chinese were the undisputed leaders in science and exploration. They invented gunpowder, the compass, and the printing press. They were unparalleled in military power and technology. Meanwhile, medieval Europe was wracked by religious wars and mired in inquisitions, witch trials, and superstition, and great scientists and visionaries like Giordano Bruno and Galileo were often either burned alive or placed under house arrest, their works banned. Europe, at the time, was a net importer of technology, not a source of innovation.

The Chinese emperor launched, under the command of Admiral Zheng He, the most ambitious naval expedition of all time, with twenty-eight thousand sailors on a fleet of 317 huge ships, each one five times longer than the ships of Columbus. The world would not see anything like it for another four hundred years. Not once, but seven times, from 1405 to 1433, Admiral Zheng He sailed across the known world, around Southeast Asia and past the Middle East, eventually ending up in East Africa. There are ancient woodcuts of the strange animals, like giraffes, that he brought back from his voyages of discovery being paraded before the court.

But when the emperor passed away, the new rulers decided that they had no use for exploration and discovery. They even decreed that a Chinese citizen could not own a boat. The fleet itself was left to rot or allowed to burn, and records of Admiral Zheng He's great accomplishments were suppressed. Succeeding emperors effectively cut off contact between China and the rest of the world. China turned inward, with disastrous results, eventually leading to decay, total collapse, chaos, civil war, and revolution.

I sometimes think about how easy it is for a nation to slip into complacency and ruin after decades of basking in the sun. Since science is the engine of prosperity, nations that turn their backs on science and technology eventually enter a downward spiral.

The U.S. space program similarly fell into decline. But now the political and economic circumstances are changing. A new cast of characters is taking center stage. Daring astronauts are being replaced by dashing billionaire entrepreneurs. New ideas, new energy, and new funding are driving this renaissance. But can this combination of private funds and government financing pave the way to the heavens?

> Yours is the light by which my spirit's
> born. You are my sun, my moon, and
> all my stars.
> —E. E. CUMMINGS

2 NEW GOLDEN AGE OF SPACE TRAVEL

Unlike the decline of China's naval fleet, which lasted for centuries, the U.S. manned space program is experiencing a revival after only a few decades of neglect. A variety of factors is turning the tide.

One is the influx of resources from Silicon Valley entrepreneurs. A rare combination of private funding and governmental financing is making possible a new generation of rockets. At the same time, the falling cost of space travel allows a range of projects to become feasible. Public support for space travel is also reaching a tipping point, as Americans again warm up to Hollywood movies and TV specials about space exploration.

And most important, NASA has finally regained its focus. On October 8, 2015, after years of muddle, vacillation, and indecision, NASA finally declared its long-term goal: to send astronauts to Mars. NASA even sketched out a rough set of goals for itself, beginning with returning to the moon. Rather than a final destination, though, the moon would be a stepping-stone for the more ambitious goal of reaching Mars. The once rudderless agency suddenly had a direction. Analysts hailed this decision, concluding that NASA was once again claiming the mantle of leadership in space exploration.

So first, let us discuss our nearest celestial neighbor, the moon, and then travel outward into deep space.

GOING BACK TO THE MOON

The backbone of NASA's effort to return to the moon is a combination of the Space Launch System (SLS) heavy booster rocket and the Orion space module. Both of them are orphans of President Obama's budget cuts of the early 2010s, when he canceled the Constellation program. But NASA was able to salvage the space module of the Constellation, the Orion capsule, as well as the SLS heavy booster rocket, which was still in the planning stage. Originally from entirely different missions, they were cobbled together to create NASA's basic launch system.

Currently, the SLS/Orion rocket is scheduled to conduct a manned lunar flyby in the mid-2020s.

The first thing you notice about the SLS/Orion system is that it does not look anything like its immediate predecessor, the space shuttle. It does, however, resemble the Saturn V rocket. For about forty-five years, the Saturn V rocket has been a museum piece. But in some sense, it is now being resurrected as the SLS booster rocket. Seeing the SLS/Orion brings on a sense of déjà vu.

The SLS can carry a payload of 130 tons. It is also 322 feet tall, comparable to the Saturn V rocket. The astronauts, instead of sitting in a ship on the side of the booster rocket like they did on the space shuttle, are in a capsule perched directly on top of the booster rocket, like the Apollo spacecraft was on the Saturn V. The SLS/Orion, unlike the space shuttle, is dedicated to carrying mainly astronauts and not cargo. In addition, the SLS/Orion is not designed to merely reach near-Earth orbit. Instead, like the Saturn V, it is designed to attain Earth escape velocity.

The Orion capsule is designed to carry four to six crew members, while the Apollo capsule of the Saturn V only held three. Like the Apollo capsule, the Orion capsule is cramped inside. It is sixteen feet in diameter and eleven feet tall and weighs fifty-seven thousand pounds. (Since space is at a premium, astronauts have historically been small people. Yuri Gagarin, for example, was only five feet two inches tall.)

And unlike the Saturn V rocket, which was specifically designed to go the moon, the SLS rocket can take you almost anywhere—to the moon, the asteroids, and even Mars.

We also have the billionaires who are tired of the lumbering pace of NASA bureaucrats and want to send astronauts to the moon and even Mars relatively soon. These young entrepreneurs were lured by former President Obama's proposal to have private enterprise take over the manned space program.

Defenders of NASA claim that its cautious pace is due to NASA's safety concerns. In the wake of the two space shuttle disasters, congressional hearings almost caused the space program to shut down entirely amid strong public disapproval. Another disaster of that scope could put an end to the program. Also, they point out that in the 1990s, NASA tried adopting the mantra "Faster, Better, Cheaper." However, when the Mars Observer was lost in 1993 due to a rup-

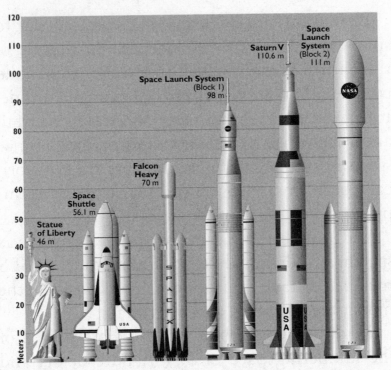

This lineup compares the original Saturn V rocket, which took our astronauts to the moon and the Space Shuttle, with other booster rockets being tested.

tured fuel tank just as it was about to orbit Mars, many thought that NASA might have rushed the mission, and the "Faster, Better, Cheaper" slogan was quietly dropped.

So one has to strike a delicate balance between the hotheads who want an accelerated pace and the bureaucrats who are gun-shy about safety and the cost of failure.

Nevertheless, two billionaires have taken the

lead in fast-tracking the space program: Jeff Bezos, founder of Amazon and owner of the **Washington Post,** and Elon Musk, founder of PayPal, Tesla, and SpaceX.

The press is already dubbing it the "battle of the billionaires."

Both Bezos and Musk would like to shift humanity into outer space. While Musk is taking the long view and setting his sights on Mars, Bezos has a more immediate vision of going to the moon.

TO THE MOON

People from all over have flocked to Florida, hoping to catch a glimpse of the first capsule that will take our astronauts to the moon. The lunar capsule will carry three astronauts on a voyage unprecedented in human history, an encounter with another celestial body. The journey to the moon will take about three days, and the astronauts will experience things never felt before, such as weightlessness. After a heroic voyage, the ship will splash down safely in the Pacific Ocean, and its passengers will be celebrated as heroes, opening up a new chapter in world history.

All the calculations have been done using Newton's laws, ensuring a precise voyage. But there is one problem. It's actually a tale written by Jules

Verne, in his prophetic novel **From the Earth to the Moon,** published in 1865, just after the end of the Civil War. The organizers of the moon shot are not NASA scientists but members of the Baltimore Gun Club.

What is truly remarkable is that Jules Verne, writing more than one hundred years before the first lunar landing, was able to predict so many features of the actual moon shot. He was able to correctly portray the size of the capsule, the location of the launch, and the method of landing back on Earth.

The only major flaw in his book was the use of a gigantic cannon to send the astronauts to the moon. The sudden acceleration of the gunshot would be about twenty thousand times the force of gravity, which would certainly kill anyone aboard the ship. However, before the coming of liquid fueled rockets, Verne had no other way to envision the journey.

Verne also postulated that the astronauts would become weightless, but only at one point, midway between the moon and the Earth. He did not realize that the astronauts would become weightless throughout their voyage. (Even today, commentators make mistakes about weightlessness, sometimes stating that it is caused by the absence of gravity in space. Actually, there is plenty of gravity in space, enough to whip giant planets like Jupiter around the sun. The experience of weightlessness is caused by the fact that everything falls at the same rate. So an astronaut inside the spaceship would fall at the same rate as his ship and experience the illusion that gravity has been turned off.)

Today, it is not the private fortunes of the members of the Baltimore Gun Club fueling this new space race but the checkbooks of moguls like Jeff Bezos. Instead of waiting for NASA to give him permission to build rockets and a launchpad with taxpayer dollars, he founded his own company, Blue Origin, and is building them himself, with his own pocket money.

Already, the project has gone beyond the planning stage. Blue Origin has produced its own rocket system, called New Shepard (named after Alan Shepard, the first American to go into space via a suborbital rocket). In fact, the New Shepard rocket was the first suborbital rocket in the world to successfully land back on its original launchpad, just beating out Elon Musk's Falcon rocket (which was the first reusable rocket to actually send a payload into Earth orbit).

Bezos's New Shepard rocket is only suborbital, meaning that it cannot reach a speed of eighteen thousand miles per hour and go into near-Earth orbit. It won't take us to the moon, but it may be the first American rocket to routinely offer tourists a view of space. Blue Origin recently released a video of a hypothetical journey on the New Shepard, and it looked like you were riding first class on a luxury ship. When you enter the space capsule, you are immediately impressed with how roomy it is. Far from the cramped quarters often seen in science fiction movies, there is ample room for you and five other tourists to be strapped into your lush reclining seats, where you immediately sink into black leather. You can look out of huge windows that are approximately 2.4 feet wide and 3.5 feet tall. "Every seat is a window seat, the largest windows ever in space," Bezos claims. Space travel has never been so gorgeous.

Because you are about to enter outer space, there are some precautions you must take. Two days before the trip, you fly into Van Horn, Texas, where Blue Origin has its launch facility. There you meet your fellow tourists and hear short talks by the crew. Since the voyage is completely automated, the crew members do not ride along with the tourists.

Your instructor explains that the entire trip will take eleven minutes as you soar vertically, sixty-two miles straight up, reaching the boundary between the atmosphere and outer space. Outside, the sky will turn dark purple and then inky black. Once the capsule reaches outer space, you can unbuckle your seat and experience four minutes of weightlessness. You will then float like an acrobat, free of the earthly constraints of gravity.

Some people get sick and vomit while experiencing weightlessness, but this won't be a problem, the instructor claims, since the trip is so short.

(For training astronauts, NASA employs the "vomit comet," which is a KC-135 airplane that can simulate weightlessness. The vomit comet rises steeply, suddenly shuts off its engines for about thirty seconds, and then falls back down. The astronauts are now like a rock thrown in the air—they are in free fall. When the airplane turns on its engines, the astronauts fall back to the floor. This process is repeated for several hours.)

At the end of the New Shepard trip, the capsule releases parachutes and then gently lands back on the ground using its own rockets. There is no need for a splashdown in the ocean. And unlike the space shuttle, it has a safety system so that you can be ejected from the rocket if there is a misfire during launch. (The space shuttle **Challenger** did not have such an ejection system, and seven astronauts died.)

Blue Origin has not yet released the price tag for this suborbital trip into space, but analysts think initially it could be in the neighborhood of $200,000 per passenger. This is the price of a trip on a rival suborbital rocket being developed by Richard Branson, another billionaire who has made his mark in the annals of space exploration. Branson is the founder of Virgin Atlantic Airways and Virgin Galactic, and he is backing the efforts of aerospace engineer Burt Rutan. In 2004, Rutan's SpaceShipOne made headlines when it won the $10 million Ansari XPRIZE. SpaceShipOne was able to reach the boundary of the atmosphere seventy miles above the Earth. Although SpaceShipTwo suffered a fatal accident in 2014 when flying above the Mojave Desert, Branson plans to continue testing the rocket and make space tourism a reality. Time will tell which rocket system will succeed commercially. But it seems clear that space tourism is here to stay.

Bezos is producing yet another rocket that will send people into Earth orbit. It is the New Glenn rocket, named after astronaut John Glenn, the first American to orbit the Earth. The rocket will have up to three stages, stand 313 feet tall, and generate 3.8 million pounds of thrust. Although the New Glenn rocket is still being designed, Bezos has dropped hints that he is planning an even more advanced rocket, to be called the New Armstrong, which may go beyond Earth orbit and all the way to the moon.

When he was a child, Bezos dreamed of going into outer space, mainly with the crew of the **Enterprise** on **Star Trek.** He would participate in plays based on the TV series, taking on the roles of Spock, Captain Kirk, and even the computer.

Upon his high school graduation, a time when most teenagers might fantasize about their first car or the senior prom, he laid out a visionary plan for the next century. He said he wanted to "build space hotels, amusement parks, yachts and colonies for two or three million people orbiting around the earth."

"The whole idea is to preserve the Earth . . . The goal [is] to be able to evacuate humans. The planet would become a park," he wrote. As Bezos saw it, the polluting industrial output of the planet could eventually be moved into space.

To put his money where his mouth is, as an adult, he founded the company Blue Origin to build the rockets of the future. The name of his rocket company refers to the planet Earth, which can be seen as a blue sphere from outer space. The aim is "to open up space travel to paying customers. The vision for Blue is pretty simple," he says. "We want to see millions of people living and working in space. That's going to take a long time but I think it's a worthwhile goal."

In 2017, he announced a short-term plan for Blue Origin to set up a delivery system for the moon. He envisions a vast operation that, just as Amazon rapidly ships out a variety of products at the click of a button, could deliver machinery, building supplies, and goods and services to the moon. Once considered a lonely outpost in space, the moon would become a bustling industrial and commercial hub, with permanent manned bases and manufacturing.

This loose talk about cities on the moon might normally be dismissed as the ravings of an eccentric. But when it comes from one of the richest people on Earth, who has the ear of the president, Congress, and the editors of the **Washington Post,** one takes it all quite seriously.

PERMANENT MOON BASE

To help pay for these ambitious projects, astronomers have looked into the physics and economics of mining the moon and noted at least three potential resources worth exploiting.

In the 1990s, an unexpected discovery caught scientists by surprise: the presence of large quantities of ice in the southern hemisphere of the moon. There, in the shadows of large mountain ranges and craters, is a perpetual darkness that is below freezing. The origin of this ice is probably cometary impacts in the early history of the solar system. Comets are mainly made of ice, dust, and rock, so any comet that strikes the moon in one of these shadows might leave a deposit of water and ice. The water, in turn, can be turned into oxygen and hydrogen (which happen to be the principal components of rocket fuel). This could turn the moon into a cosmic gas station. The water could also be purified for drinking purposes or used to create small-scale agricultural farms.

In fact, another group of Silicon Valley entrepreneurs has created a company called Moon Express to begin the process of mining ice from the moon. It is the first company ever to get permission from the government to begin this commercial enterprise. The preliminary target for Moon Express is, however, more modest. The company will begin by putting a lunar rover on the moon that will systematically search for the presence of ice deposits. The company has already raised enough money through private funding to proceed with this mission. With the financing in place, all systems are go.

Scientists have analyzed the moon rocks brought back by the Apollo astronauts and believe there may be other economically significant elements on the moon. Rare earth elements are crucial for the electronics industry but are mostly found in China. (Rare earths are located everywhere in small quantities, but the Chinese rare earth industry makes up 97 percent of the world trade. China has roughly 30 percent of the world's reserves.) A few years ago, an international trade war almost erupted when Chinese suppliers abruptly raised prices on these key elements, and the world suddenly realized that China had a near monopoly. It is estimated that the supply will begin to be depleted in the coming decades, making it urgent to find alternate sources. Rare earths have been found in moon rocks, so one day it may be cost-effective to extract them from the moon. Platinum is another important element for the electronics industry, and the presence of platinum-like minerals, perhaps left over from ancient asteroid impacts, has also been detected on the moon.

Finally, there is the possibility of finding helium-3, which is useful in fusion reactions. When hydrogen atoms are combined at the extremely high temperatures found in these reactions, the hydrogen nuclei fuse, creating helium, plus large amounts of energy and heat. This excess energy is useful to power machines. However, this process also produces copious quantities of neutrons, which are dangerous. The advantage of the fusion process involving helium-3 is that it instead releases an excess proton, which can be handled more easily and deflected by electromagnetic fields. Fusion reactors are still highly experimental, and so far, none exist on Earth. But if they are successfully developed, helium-3 could be mined from the moon to supply fuel for the fusion reactors of the future.

But this also raises a tricky point: Is it legal to mine the moon? Or to stake a claim there?

In 1967, the United States, Soviet Union, and many other nations signed the Outer Space Treaty, which banned nations from claiming ownership of celestial bodies like the moon. It banned nuclear weapons from Earth orbit and from being placed on the moon or elsewhere in space. The testing of these weapons was also prohibited. The Outer Space Treaty, the first and only one of its kind, holds to this day.

However, the treaty said nothing about private ownership of land or the use of the moon for commercial activities, probably because those who drafted it didn't believe private individuals would ever be able to reach the moon. But these matters must be addressed soon, especially now that the price of space travel is dropping and billionaires want to commercialize outer space.

The Chinese have announced that they will put their astronauts on the moon by 2025. If they plant their flag, it will largely be symbolic. But what happens if some private developer stakes a claim to the moon after arriving on his or her private spaceship?

Once these technical and political issues are settled, the next question is, What will it be like to actually live on the moon?

LIVING ON THE MOON

Our original astronauts only spent brief periods of time on the moon, usually a few days. To create the first manned outposts, future astronauts would have to spend extended time there. They would need to adjust to lunar conditions, which, as you can imagine, are quite different from the Earth's.

One factor that limits how long our astronauts can stay on the moon is the availability of food, water, and air, since they would exhaust the supplies that they carry with them within a matter of weeks. In the beginning, everything would have to be shipped from the Earth. Unmanned lunar probes would have to be sent every few weeks to resupply the station. These shipments would become a lifeline for the astronauts, so any accident involving them could present an emergency. Once a moon base is constructed, even a temporary one, one of the first endeavors for the astronauts might be to create oxygen for breathing and for growing their own food. There are a number of chemical reactions that can produce oxygen, and the presence of water creates a ready supply. And this water could also be used in hydroponic gardens to grow crops.

Fortunately, communication with the Earth would not pose much of a problem, since it only takes a little more than a second for a radio signal to reach the Earth from the moon. Apart from a slight delay, astronauts would be able to use their cell phones and the internet like they do on Earth, so they could be in constant contact with their loved ones and receive the latest news.

Initially, our astronauts would have to live inside the space capsule. When they venture out, the first order of business would be to unfurl large solar panels to harvest energy. Since one lunar day corresponds to one Earth month, any site on the moon has two weeks of daylight followed by two weeks of darkness. So they would need large banks of batteries to store the electrical energy harvested during the two-week "day" for use during the long "night" that follows.

Once on the moon, the astronauts might want to travel to the poles for several reasons. There are peaks in the polar regions where the sun never sets, so a solar farm with thousands of solar panels could create a steady supply of energy without interruption. The astronauts might also take advantage of the deposits of ice in the shadows of large mountain ranges and craters at the poles. It is estimated that six hundred million metric tons of ice may be found in the northern polar region, in a layer that is several yards thick. Once mining operations begin, much of this ice could be harvested and purified for drinking purposes, as well as for oxygen. It is also possible to mine the soil of the moon, which contains a surprising amount of oxygen. In fact, there are about one hundred pounds of oxygen for every one thousand pounds of lunar soil.

The astronauts would have to adjust to the lower gravity on the moon. According to Newton's theory of gravity, the amount of gravity on any planet is related to its mass. The moon's gravity is one-sixth that of Earth's.

This means that moving heavy machinery would be much easier on the moon. And the escape velocity is much lower, so rockets could both land and take off from the moon rather easily. In the future, a busy spaceport on the moon is a distinct possibility.

But our astronauts would have to relearn simple movements, such as walking. Apollo astronauts realized that walking on the moon was quite awkward. They found that the fastest way to maneuver was to hop. Because of the moon's lower gravity, you can hop much farther than taking a step, and it's easier to control your motion.

Another issue to contend with is radiation. For missions lasting a few days, it does not pose a major problem. But if the astronauts spend months on the moon, they could accumulate enough exposure to seriously increase their risk of getting cancer. (Simple medical problems could easily escalate into life-threatening situations on the moon. All of the astronauts would have to have first aid training, and a few of them would probably be medical doctors. If, for example, an astronaut has a heart attack or appendicitis on the moon, most likely the doctor would set up a tele-conferencing session with specialists on Earth, who would perhaps perform surgery by remote control. Robots could be brought in to do various forms of microsurgery, guided by skillful hands back on Earth.) The astronauts would need daily "weather reports" from astronomers monitoring solar activity. Instead of indicating upcoming thunderstorms, these weather reports would give warning of huge solar flares that send hot plumes of radiation into space. If there is a giant eruption on the sun, the astronauts could be signaled to seek cover. Once the warning is given, astronauts would have several hours before a deadly rain of charged subatomic particles hits the base.

One way to create shelter from radiation might be to dig an underground base within a lava tube on the moon. These tubes, remnants of ancient volcanoes, can be huge, up to a thousand feet across, and would provide adequate protection from radiation from the sun and outer space.

Once the astronauts have erected the temporary shelter, large shipments of machinery and supplies would have to be sent from the Earth to begin construction of the permanent moon base. Shipping prefabricated materials and inflatable items could speed up this process. (In the movie **2001**, astronauts live in huge, modern underground lunar bases, which contain landing platforms for rockets and serve as headquarters to coordinate lunar mining operations. Our first lunar headquarters may not be as comprehensive, but the vision presented in the movie may be realized before too long.)

In building these underground bases, inevitably you will want the ability to manufacture and repair machine parts. Although large equipment such as bulldozers and cranes would have to be sent from the Earth, 3-D printers could fabricate small plastic machine parts on-site.

Ideally, factories would be established to forge metal. But building a blast furnace is impossible, since there is no air to feed the furnace. However, experiments have shown that lunar soil, when heated by microwaves, can be melted and fused to make rock-hard ceramic bricks, which could be the basic building blocks for the entire moon base. In principle, all of the infrastructure could be made of this material, which can be harvested directly from the soil.

LUNAR RECREATION/ENTERTAINMENT

Lastly, there has to be a source of entertainment for the astronauts, a way to blow off steam and relax. When Apollo 14 landed on the moon in 1971, NASA officials did not know that commander Alan Shepard had secretly smuggled a six-iron golf club into the space capsule. They were surprised when he proceeded to take out the club and hit a golf ball two hundred yards on the lunar surface. This was the first and only time someone engaged in a sports activity on another celestial body. (A replica of the golf club is now in the Smithsonian National Air and Space Museum in Washington, D.C.) Lunar sports would pose a particular challenge due to the lack of air and low gravity. But they will also give rise to some remarkable feats.

On Apollo 15, 16, and 17, our astronauts rode the Lunar Roving Vehicles over the dusty surface, each traveling between seventeen and twenty-two miles. Not only was this a valuable scientific mission, it was also a thrilling expedition as they looked out at majestic craters and mountain ranges, knowing that they were the first people ever to see these stunning sights. In the future, riding dune buggies will not only accelerate the survey of the lunar surface, the installation of solar panels, and the construction of the first lunar station, it will serve as a type of recreation. It may perhaps even make possible the first races on the moon.

Lunar tourism and exploration could become popular recreational activities as people discover the wonders of an alien landscape. Given the low gravity, hikers would be able to trek over long distances without tiring. Mountain climbers would be able to rappel down steep mountainsides with little effort. And from the top of craters and mountain ranges, they would have an unprecedented panorama of the lunar landscape, literally untouched for billions of years. Spelunkers who love to explore caves will be excited to investigate the network of gigantic lava tubes crisscrossing the moon. On the Earth, caves were carved out by underground rivers and contain evidence of ancient water flows in the form of stalactites and stalagmites. But on the moon, there are no appreciable deposits of liquid water. The moon's caves were instead carved out of the rock by molten lava flows. They would look completely different from the caverns we see on Earth.

WHERE DID THE MOON COME FROM?

Once mining operations successfully exploit the resources found on the surface of the moon, we will inevitably turn our eye to the riches that may lie deep within it. Uncovering them would change the economic landscape, as the accidental and unexpected discovery of oil on Earth did. But what is the interior of the moon like? To answer this, we have to consider the question, Where did the moon come from?

The origin of the moon has fascinated humanity for millennia. Because the moon rules the night, it has often been associated with darkness or madness. The word **lunatic** comes from **luna,** the Latin word for moon.

Ancient mariners were fascinated by the connection between the moon, the tides, and the sun, and correctly ascertained that there is a close correlation between all three.

The ancients noticed another curious fact: you only see one side of the moon. Think of all the times you've looked at the moon, and you realize you're always seeing the same face.

It was Isaac Newton who finally put all the puzzle pieces together. He calculated that the tides are caused by the gravitational pull of the moon and the sun on the Earth's oceans. His theory indicated that the Earth creates tidal effects on the moon as well. Since the moon is made of rock and has no oceans, it is actually being squeezed by the Earth, and this force causes it to bulge slightly. At one time, it tumbled in its orbit around the Earth. Eventually, this tumbling slowed down, until the spinning of the moon was locked to the Earth, so that one side of it always faced us. This is called tidal locking, and it happens throughout the solar system, including for moons of Jupiter and Saturn.

Using Newton's laws, you can also determine that tidal forces are causing the moon to slowly spiral away from the Earth. Its orbital radius increases by about 1.6 inches per year. This small effect can be measured by shooting laser beams to the moon—our astronauts left a mirror behind to help with this experiment—and then calculating the time it takes for the beams to bounce back to Earth. The round trip takes only about two seconds, but this number is gradually increasing. If the moon is spiraling away, then, by running the videotape backward, we can estimate its past orbit.

A quick calculation shows that the moon separated from the Earth billions of years ago. And modern evidence indicates that 4.5 billion years ago, not long after the Earth was formed, there was a cosmic impact between the Earth and a large asteroid of some sort. This asteroid, which we call Theia, was about the size of Mars. Computer simulations have given us dramatic insight into this explosion, which gouged out a huge chunk of the Earth and propelled it into space. But because the impact was more of a glancing blow than a direct strike, it didn't take much of the interior iron core of the Earth. As a result, the moon, while it does contain some iron, has no significant magnetic field because it lacks a molten iron core.

After the collision, the Earth resembled a Pac-Man, with a huge pie-shaped piece carved out. But because of the attractive nature of gravity, eventually both the moon and Earth condensed into spheres again.

Evidence of the impact theory was provided by the astronauts who brought 842 pounds of rock back from their historic trips to the moon. Astronomers discovered that the moon and the Earth are made of almost the same chemicals, including silicon, oxygen, and iron. By contrast, random analysis of rocks from the asteroid belt shows that their composition is quite different from that of the Earth's.

I had my own encounter with moon rocks when I was a graduate student in theoretical physics at the Berkeley Radiation Laboratory. I had a chance to view one under a powerful microscope. I was surprised by what I saw. There were tiny craters caused by micrometeors that had impacted the moon billions of years ago. Then, looking more closely, I saw craters inside these craters. And even smaller craters inside those. This chain of craters-inside-craters would be impossible in Earth rock, since these micrometeors would have vaporized while going through the atmosphere. But they could hit the lunar surface because the moon has no atmosphere. (This also means that micrometeors could be a problem for astronauts on the moon.)

Since the composition of the moon is so similar to Earth's, the truth may be that mining the interior of the moon is only useful if you are building cities on the moon. It is probably too expensive to bring moon rock back to Earth if it only offers what we already have. But lunar material could be immensely valuable for creating a local infrastructure of buildings, roads, and highways on the moon.

WALKING ON THE MOON

What would happen if you took off your space suit on the moon? Without air, you would suffocate, but there is something even more disturbing: your blood would boil.

At sea level, water boils at 212 degrees Fahrenheit or 100 degrees Celsius. The boiling point of water drops as atmospheric pressure drops. As a child, I had a vivid demonstration of this principle one day while camping in the mountains. We were frying eggs in a pan over a fire. The eggs, sizzling away in the pan, looked delicious. But when I ate them, I almost threw up. They tasted awful. Then it was pointed out to me that as you climb up a mountain, the atmospheric pressure begins to drop, and the boiling point of water decreases. Although the eggs bubbled and appeared to be fried, they were never completely cooked. The bubbling egg wasn't so hot at all.

I had another encounter with this fact when celebrating Christmas as a child. At our house, we had old-fashioned Christmas lights consisting of thin tubes of water placed vertically on top of each electrical fixture. When we turned them on, they were gorgeous. The brightly colored water in the tubes began to boil in various colors. Then I did something foolish. I grabbed the tubes of boiling water with my bare fingers. I immediately expected to feel the intense heat of boiling water, but I felt almost nothing. Years later, I realized what had happened. Inside the tube was a partial vacuum. As a consequence, the boiling point of water dropped, so even the heat of a small electrical fixture could make the liquid boil, but the boiling water wasn't hot at all.

Our astronauts will encounter the same physics if they ever have a leak in their space suits in space or on the moon. As the air leaves the suit, the pressure inside drops and the boiling point of water also drops. Eventually, the blood in the astronaut's body will begin to boil.

Sitting in our chair here on Earth, we forget that we have almost fifteen pounds of air pressure pushing down on every square inch of our skin because there is a huge column of air sitting right above us. Why aren't we crushed? Because we have fifteen pounds of pressure pushing out from inside our body. There is a balance. But if we go to the moon, the fifteen pounds of pressure beating down on us from the atmosphere disappears. Then we only have the fifteen pounds of pressure pushing outward.

In other words, taking off your space suit on the moon could be a very unpleasant experience. Best to keep it on at all times.

What might a permanent moon base look like? Unfortunately, NASA has not issued any formal blueprints, so all we have are the imaginations of science fiction authors and Hollywood script-writers as rough guides. But once a lunar base is constructed, we would endeavor to make it totally self-sustaining. Such a system would vastly lower costs. But it would require a great deal of infrastructure: factories to create buildings, large greenhouses for food, chemical plants to create oxygen, and huge solar banks for energy. To pay for all of this, one would need a source of income. Since the moon is largely made of the same material as the Earth, we may need to look beyond it for a revenue stream. That is why Silicon Valley entrepreneurs have already set their sights on the asteroids. There are millions of asteroids in space, and they may be the home of untold riches.

Killer asteroids are nature's way of asking, "How's that space program coming along?"
—ANONYMOUS

3 MINING THE HEAVENS

Thomas Jefferson was deeply disturbed.

He had just signed over $15 million to Napoleon, a princely sum in 1803, the most controversial and costly decision of his career as president. He had doubled the size of the United States. The country would now extend all the way to the Rocky Mountains. The Louisiana Purchase would go down as one of the biggest successes, or failures, of his presidency.

Looking at the map, with its huge expanse of totally uncharted territory, he wondered if he would regret his decision.

Eventually, he would send Meriwether Lewis and William Clark on a mission to explore what he had bought. Was it a wilderness paradise waiting to be colonized or a desolate wasteland?

Privately, he acknowledged that in any event, it might take another thousand years to settle such a vast stretch of land.

A few decades later, something happened that changed everything. In 1848, gold was discovered at Sutter's Mill in California. The news was electrifying. More than three hundred thousand people flooded into this wilderness to seek riches. Ships from all over began to line up at San Francisco harbor. Its economy exploded in size. The next year, California applied for statehood.

Farmers, ranchers, and businessmen followed, making possible the formation of some of the first great cities of the West. In 1869, the railroad came to California, connecting it to the rest of the United States and supporting an infrastructure of transportation and commerce that led to rapid population growth in the region. The mantra for the nineteenth century was, "Go west, young man." The Gold Rush, for all its excesses, helped to open up the West for settlement and make all this happen.

Today, some are wondering whether the mining of the asteroid belt could create another Gold Rush in outer space. Already private entrepreneurs have expressed an interest in exploring this region and its untold riches, and NASA has funded several missions with the goal of bringing an asteroid back to Earth.

Could the next great expansion be in the asteroid belt? And if so, how might we incorporate and sustain this new space economy? One can envision a potential analogy between the agricultural supply chain of the nineteenth-century Wild West and a future supply chain involving the asteroids. In the 1800s, teams of cowboys would herd cattle from ranches in the Southwest almost a thousand miles toward cities like Chicago. There, the beef would be processed and sent farther east by train to satisfy demand in urban areas. In the same way that these early cattle drives connected the Southwest to the Northeast, perhaps an economy could arise connecting the asteroid belt to the moon and the Earth. The moon would be like the Chicago of the future, processing valuable minerals from the asteroid belt and shipping them on to Earth.

ORIGIN OF THE ASTEROID BELT

Before we delve further into the details of asteroid mining, it may be helpful to clarify a few
terms that are often confused with one another:
meteor, meteorite, asteroid, and **comet**. A **meteor** is a piece of rock that burns up in the atmosphere as it streaks across the sky. The tails
of meteors, which point away from the direction
of motion, are caused by air friction. On a clear
night, you might see a meteor every few minutes
simply by gazing upward.

A rock that actually lands on Earth is called a
meteorite.

Asteroids are rocky debris in the solar system.
Most of them are contained in the asteroid belt
and are remnants of a failed planet between Mars
and Jupiter. If you were to add up the masses
of all the known asteroids, the sum would only
amount to 4 percent of the mass of the moon.
However, the majority of these objects have not
yet been detected by us, and there are potentially
billions of them. For the most part, asteroids
remain in stable orbits in the asteroid belt, but
occasionally one strays and hits the Earth's atmosphere and burns up as a meteor.

A **comet** is a piece of ice and rock that origi-
nates far beyond the orbit of the Earth. While asteroids lie within the solar system, many comets

actually orbit in the outer fringes of the solar system, in the Kuiper Belt, or even outside the solar system itself, in the Oort Cloud. The comets we see in the night sky are those whose orbit or trajectory has brought them near the sun. When comets approach the sun, solar wind pushes particles of ice and dust away from the comet, resulting in tails that point away from the sun, not away from the direction of motion.

Over the years, a picture has emerged of how our solar system was formed. About five billion years ago, our sun was a slowly spinning gigantic cloud, mainly made of hydrogen and helium gas and dust. It was several light-years across (a light-year is the distance light travels in one year, or about six trillion miles). Because of its large mass, it was gradually compressed by gravity. As it shrank in size, it rotated faster and faster, just as skaters spin faster when they bring their arms in. Eventually the cloud condensed into a rapidly rotating disc with the sun at its center. The surrounding disc of gas and dust began to form protoplanets, which got larger as they continued to absorb material. This process explains why all the planets revolve around the sun in the same direction, in the same plane.

It is believed that one of these protoplanets got too close to Jupiter, the largest of the planets, and was ripped apart by its enormous gravity, thereby forming the asteroid belt. Another theory suggests that the collision of two protoplanets may have resulted in the asteroid belt.

The solar system can be pictured as four belts orbiting the sun: the innermost belt is made up of the rocky planets, which include Mercury, Venus, Earth, and Mars; next is the asteroid belt; beyond that is the gas giant belt, consisting of Jupiter, Saturn, Uranus, and Neptune; and finally the comet belt, also called the Kuiper Belt. And outside these four belts, we have a spherical cloud of comets surrounding the solar system called the Oort Cloud.

Water, a simple molecule, was a common substance in the early solar system but occurred in different forms depending on its distance from the sun. Close to the sun, where water would boil and turn to steam, we find the planets Mercury and Venus. The Earth is farther out, so that water can exist in liquid form. (This is sometimes called the "Goldilocks zone," where the temperature is right for liquid water to exist.) Beyond that, water turns to ice. So Mars and the planets and comets beyond that mainly have water in frozen form.

MINING THE ASTEROIDS

Understanding the origin of asteroids and therefore their composition will be crucial for mining operations.

The idea of mining asteroids is not as preposterous as it might seem. We actually know a considerable amount about their makeup, because some of them hit the Earth. They consist of iron, nickel, carbon, and cobalt, and they also contain significant quantities of rare earths and valuable metals such as platinum, palladium, rhodium, ruthenium, iridium, and osmium. These elements are found naturally on Earth, but they are rare and very expensive. As the supply of these resources on Earth is exhausted in the coming decades, it will become economical to mine them in the asteroid belt. And if an asteroid is nudged so that it orbits the moon, it can be mined at will.

In 2012, a group of entrepreneurs established a company called Planetary Resources to extract valuable minerals from asteroids and bring them back to Earth. This ambitious and potentially highly lucrative plan was backed by some of the biggest players in Silicon Valley, including Larry Page, CEO of Google's parent company, Alphabet, Inc., and executive chairman Eric Schmidt, as well as Oscar-winning director James Cameron.

Asteroids, in some sense, are like flying gold mines in outer space. For example, in July 2015, one came within a million miles of Earth, or about four times the distance from the Earth to the moon. It was about nine hundred meters (or about three thousand feet) across and was estimated to contain ninety million tons of platinum in its core, worth $5.4 trillion. Planetary Resources estimates that the platinum within a mere thirty-meter asteroid could be worth $25 to $50 billion. The company has gone so far as to make a list of small nearby asteroids that are ripe for the taking. If any one of these were to be successfully brought back to Earth, it would contain a mother lode of minerals that would pay back its investors manyfold.

Out of the sixteen thousand or so asteroids considered near-Earth objects (those whose orbits cross the Earth's path), astronomers identified their own roster of twelve that would make ideal candidates for retrieval. Calculations have shown that these twelve, each between ten and seventy feet across, can be coaxed into lunar or Earth orbit with a gentle shift in their trajectories.

But there are many others out there. In January 2017, a new asteroid was unexpectedly detected by astronomers just hours before it whizzed by. It passed a mere thirty-two thousand miles from Earth (or 13 percent of the distance from the Earth to the moon). Fortunately, it was only twenty feet across and would not have caused significant damage if it had hit us. However, it did provide further confirmation of the great number of asteroids that drift past the Earth, most of them undetected.

EXPLORING THE ASTEROIDS

Asteroids are so important that NASA has targeted the exploration of them as the first step toward a Mars mission. In 2012, a few months after Planetary Resources unveiled its plan at a press conference, NASA announced the Robotic Asteroid Prospector project, which will analyze the feasibility of mining them. Then, in the fall of 2016, NASA launched a billion-dollar probe, called OSIRIS-REx, to meet Bennu, an asteroid measuring sixteen hundred feet across that will pass the Earth in 2135. By 2018, the probe will circle Bennu, land on it, and then bring back between two and seventy ounces of rock to Earth for analysis. This plan is not without risk, as NASA fears that even slight perturbations in the orbit of Bennu might cause it to hit the Earth on its next pass. (If it does strike the Earth, it would do so with the force of a thousand Hiroshima bombs.) This mission, however, could provide invaluable experience in intercepting and analyzing objects in space.

NASA is also developing the Asteroid Redirect Mission (ARM), which aims to actually retrieve asteroid boulders from space. Funding is not guaranteed, but the hope is that the mission could open up a new source of revenue for the space program. The ARM has two stages.

First, an unmanned probe would be sent into deep space to intercept an asteroid that has been carefully evaluated by Earth-based telescopes. After conducting a detailed survey of the surface, it would land and use pincerlike hooks to grab onto a large boulder. The probe then would blast off and head to the moon, dragging the object by a tether.

At that point, a manned mission would leave Earth, using the SLS rocket with the Orion module. The module would dock with the robotic probe as they both orbit the moon. Astronauts would leave the Orion, access the probe, and extract samples for analysis. Finally, the Orion space module would separate from the robotic probe and head back to Earth, splashing down in the ocean.

One possible complication to this mission is that we don't yet know much about the physical structure of asteroids. They may be solid, or they may be a collection of smaller rocks held together by gravity, in which case they would fall apart if we tried to land on them. For this reason, further investigation is needed before this mission can proceed.

One notable physical feature of asteroids is their highly irregular shape. They often look like deformed potatoes, and the smaller they are, the more irregular they tend to be.

This, in turn, raises a question that children often ask: Why are stars, the sun, and the planets all round? Why can't stars and planets be shaped like cubes or pyramids? While small asteroids have little mass and hence little gravity to reshape them, large objects like planets and stars have huge gravitational fields. This gravity is uniform and attractive and hence will compress an irregularly shaped object into a sphere. So the planets, billions of years ago, were not necessarily round, but over time the attractive force of gravity compressed them into smooth spheres.

Another question often raised by children is why space probes aren't destroyed when they go through the asteroid belt. In the movie **Star Wars**, our heroes are almost hit by the huge chunks of rock flying around. While the Hollywood portrayal is thrilling, fortunately, it does not truly represent the density of the asteroid belt, which is mainly an empty vacuum with occasional rocks passing by. Future miners and settlers who brave outer space in search of new lands will, for the most part, find the asteroid belt relatively easy to navigate.

If these stages of asteroid exploration proceed according to plan, the final goal will be to create a permanent station to maintain, resupply, and support future missions. Ceres, the largest of the objects in the belt, might make an ideal base of operations. Ceres (whose name comes from the Greek goddess of agriculture, which also gives us the word **cereal**) was recently reclassified as a dwarf planet, like Pluto, and is thought to be an object that never quite accumulated enough matter to rival its planetary neighbors. For a celestial object, it is small, about a quarter of the size of the moon, with no atmosphere and little gravity. However, for an asteroid, it is huge; it is about 580 miles across, or roughly the size of Texas, and contains one-third of the total mass of the entire asteroid belt. Given its weak gravity, it might make an ideal space station, as rockets would easily be able to land and leave the asteroid, which are important factors in building a spaceport.

NASA's Dawn Mission, launched in 2007 and orbiting around Ceres since 2015, revealed a spherical but heavily cratered mass, made primarily of ice and rock. It is theorized that many asteroids, like Ceres, contain ice, which could be processed to extract hydrogen and oxygen for fuel. Recently, using NASA's Infrared Telescope Facility, scientists observed that the asteroid 24 Themis was completely covered in ice, with traces of organic chemicals on the surface. These findings add validity to the conjecture that asteroids and comets may have brought some of the original water and amino acids to the Earth billions of years ago.

Because asteroids are small compared to moons and planets, they probably will not evolve into permanent cities for colonists. To create a stable community in the asteroid belt would be difficult. For the most part, there is no air to breathe, water to drink, energy to consume, or soil in which to grow food, and there is no gravity to speak of. Thus, asteroids will more likely become temporary quarters for miners and robots.

But they may prove to be an essential staging area for the main event, a manned mission to Mars.

Mars is there, waiting to be reached.
—BUZZ ALDRIN

I would like to die on Mars—just not
on impact.
—ELON MUSK

4 MARS OR BUST

Elon Musk is a bit of a maverick, an entrepreneur with a cosmic mission: to build the rockets that one day will take us to Mars. Tsiolkovsky, Goddard, and von Braun all dreamed of going to Mars, but Musk may actually do it. In the process, he is breaking all the rules of the game.

He fell in love with the space program as a child growing up in South Africa and even built a rocket on his own. His father, an engineer, encouraged his interests. Early on, Musk concluded that the risk of human extinction could only be avoided by reaching for the stars. And so he decided that one of his goals would be "making life multiplanetary," a theme that has guided his entire career.

In addition to rocketry, he was compelled by two other passions, computers and business. He was programming at the age of ten and sold his first video game, called **Blaster,** for five hundred dollars at the age of twelve. He was restless and hoped one day to move to America. When he was seventeen, he emigrated to Canada on his own. By the time he received his bachelor's degree in physics from the University of Pennsylvania, he was torn between two possible careers. One path led to the life of a physicist or engineer, designing rockets or other high-tech devices. The other led to business and the use of his computer skills to amass a fortune, which would give him the means to bankroll his vision privately.

The dilemma came to a head when he began his Ph.D. studies at Stanford University in 1995 in applied physics. After spending just two days in the program, he abruptly dropped out and plunged into the world of internet start-ups. He borrowed $28,000 and founded a software company that produced an online city guide for the newspaper publishing industry. He sold it to

Compaq for $341 million four years later. He netted $22 million from that sale and immediately plowed the profits into a new company called X.com, which would evolve into PayPal. In 2002, eBay bought PayPal for $1.5 billion, from which Musk received $165 million.

Flush with cash, he harnessed these funds to fulfill his dreams, creating SpaceX and Tesla Motors. At one point, he invested 90 percent of his entire net worth in his two companies. Unlike other aerospace companies, which build rockets based on known technology, SpaceX pioneered a revolutionary design for a reusable rocket. Musk's goal was to reduce the cost of space travel by a factor of ten by reusing the booster, which is normally discarded after each launch.

Almost from scratch, Musk developed the Falcon (named after the **Millennium Falcon** from **Star Wars**), which would boost a space module called the Dragon (named after the song "Puff, the Magic Dragon") into outer space. In 2012, SpaceX's Falcon rocket made history by being the first commercial rocket to reach the International Space Station. It also became the first rocket to land successfully back on Earth after an orbital flight. His first wife, Justine Musk, says, "I like to compare him to the Terminator. He sets his program and just . . . will . . . not . . . stop."

In 2017, he scored another major victory when he successfully relaunched a used booster rocket. The previously launched rocket had landed back on its launchpad, been cleaned up and serviced, and was sent up a second time. Reusability may revolutionize the economics of space travel. Think of the used-car market. After World War II, cars were still out of reach for many people, especially GIs and the young. The used-car industry enabled average consumers to purchase cars and changed everything, including our lifestyles and social interactions. Today, in the United States, about forty million used cars are sold every year, which is 2.2 times the number of new cars sold. In the same way, Musk hopes that his Falcon rocket will transform the aerospace market and allow rocket prices to plunge. Most organizations will not care if the rocket that sends its satellite into outer space is brand-new or previously used. They will opt for the cheapest, most reliable method.

The first reusable rocket was a milestone, but Musk shocked audiences when he laid out the details of his ambitious plans to reach Mars. He expects to send an unmanned mission to Mars in 2018, followed by a manned mission by 2024, beating NASA by about a decade. His ultimate aim is to establish not just an outpost but a whole city on Mars. He imagines sending a fleet of a thousand modified Falcon rockets, each carrying one hundred colonists, to create the first settlement on the Red Planet. The keys to Musk's plan are the dramatically falling cost of space travel and new innovations. Calculations of the price of a Mars mission usually range between $400 to $500 billion, but Musk estimates that he can create and launch the Mars rocket for only $10 billion. At first, tickets to Mars would be expensive, but they would eventually drop to about $200,000 per person round trip because of the falling price of space travel. This is comparable to the $200,000 necessary to ride just seventy miles above the Earth on Virgin Galactic's SpaceShipTwo, or the $20 to $40 million estimated price of a trip to the International Space Station on a Russian rocket.

Musk's proposed rocket system was originally called the Mars Colonial Transporter, but he renamed it the Interplanetary Transport System because, as he said, "This system really gives you freedom to go anywhere you want in the greater solar system." His long-term vision is to build a network that would connect the planets as the railroad connected American cities.

Musk sees potential for collaboration with other parts of his multibillion-dollar empire. Tesla has developed an advanced version of the fully electric car, and Musk is heavily invested in solar energy, which would be the primary source of power for any Martian outpost. Therefore, Musk is in an ideal position to supply the electrical machinery and solar arrays required to advance a Mars colony.

While NASA is often painfully slow and sluggish, entrepreneurs believe they can introduce fresh, innovative ideas and techniques quickly. "There's a silly notion that failure's not an option at NASA," Musk said. "Failure is an option here [at SpaceX]. If things are not failing, you are not innovating enough."

Musk is perhaps the contemporary face of the space program: brash, fearless, and iconoclastic in addition to being innovative and smart. He is a new kind of rocket scientist, the entrepreneur-billionaire-scientist. He is often compared to Tony Stark, the alter ego of Iron Man, a suave industrialist and inventor who is at home with business tycoons and engineers alike. As a matter of fact, part of the first sequel to **Iron Man** was filmed at the SpaceX headquarters in Los Angeles, and when visitors drive up to SpaceX, they are greeted with a life-size statue of Tony Stark in his Iron Man suit. Musk even influenced a space-themed runway collection at New York Fashion Week by menswear designer Nick Graham, who explained, "They say Mars is the new black—it's incredibly top-of-the-mind trending in terms of everyone's ambitions. The idea was to show the Fall 2025 collection, based on the year Elon Musk wants to get the first people on Mars."

Musk summed up his philosophy by saying, "I really don't have any other motivation for personally accumulating assets," he said, "except to be able to make the biggest contribution I can to making life multiplanetary." Peter Diamandis of the XPRIZE has said, "There's a much bigger driver here than just profitability. [Musk's] vision is intoxicating and powerful."

NEW SPACE RACE TO MARS

All this talk about Mars, of course, was bound to stir up rivalry. The CEO of Boeing, Dennis Muilenburg, has said, "I'm convinced that the first person to step foot on Mars will arrive there riding on a Boeing rocket." It was probably no accident that he made these startling remarks one week after Musk announced his Mars plans. Musk may be grabbing all the headlines, but Boeing has a long tradition of successful space travel. It was Boeing, after all, that manufactured the booster rocket for the famed Saturn V, which took our astronauts to the moon, and Boeing currently has the contract to build the massive SLS booster rocket, the foundation of NASA's planned mission to Mars.

Supporters of NASA have pointed out that public funding was crucial for major space projects of the past, such as the Hubble Space Telescope, one of the jewels of the space program. Would private investors have funded such a risky endeavor with no hope of generating returns for stockholders? The backing of large, bureaucratic organizations may be necessary for ventures that are too expensive for private enterprise or have little hope of generating revenue.

There are advantages to each of these competing programs. Boeing's SLS, which can lift 130 metric tons into outer space, can send bigger payloads than Musk's Falcon Heavy, which can carry 64 metric tons. However, the Falcon may be more affordable. At present, SpaceX has the cheapest rates for launching satellites into space at about a thousand dollars per pound, or 10 percent of the usual rate for commercial space vehicles. Prices could drop further as SpaceX perfects its reusable rocket technology.

NASA has found itself in an enviable position, with two suitors bidding for a plum project. They can, in principle, still decide between the SLS and the Falcon Heavy. When asked about the challenge from Boeing, Musk said, "I think it's good for there to be multiple paths to Mars . . . to have multiple irons in the fire . . . You know, the more the better."

NASA spokesmen have said, "NASA applauds all those who want to take the next giant leap—and advance the journey to Mars . . . This journey will require the best and the brightest . . . At NASA, we've worked hard over the past several years to develop a sustainable Mars exploration plan, and to build a coalition of international and private sector partners to support this vision." In the end, the spirit of competition will likely prove an asset for the space program.

There is some poetic justice to this contest, however. The space program, by forcing the miniaturization of electronics, opened the doors to the computer revolution. Inspired by their childhood memories of the space program, the billionaires created by the computer revolution are coming full circle and putting some of their wealth back into space exploration.

The Europeans, Chinese, and Russians have also expressed a desire to send a manned mission to Mars in the 2040 to 2060 timeframe, but funding for these projects remains problematic. It is fairly certain, however, that the Chinese will reach the moon in 2025. Chairman Mao once lamented that China was so backward it could not launch a potato into space. Things have changed completely since then. Improving on rockets bought from Russia in the 1990s, China has already launched ten "taikonauts" into orbit and is proceeding with ambitious plans to construct a space station and develop a rocket as powerful as the Saturn V by 2020. In its various five-year plans, China is carefully following the steps pioneered by the Russians and the United States.

Even the most hopeful visionaries are fully aware that there will be a host of dangers facing astronauts on a Martian journey. Musk, when asked whether he would personally like to visit Mars, acknowledged that the probability of dying on the first trip to the planet is "quite high" and said that he would like to see his children grow up.

SPACE TRAVEL IS NO SUNDAY PICNIC

The list of the potential hazards of a manned mission to Mars is formidable.

The first is catastrophic failure. We are more than fifty years into the space age, yet the probability of a disastrous rocket accident is still around 1 percent. There are hundreds of moving parts inside a rocket, and any one of them may cause a mission to fail. The space shuttle had two horrendous accidents out of a total of 135 launches, or about a 1.5 percent failure rate. The overall fatality rate of the space program has been 3.3 percent. Of the 544 people who have ever been in space, 18 have died. Only the very courageous are willing to sit on top of a million pounds of rocket fuel to be blasted into space at twenty-five thousand miles per hour, not knowing if they are coming back.

There is also the "Mars jinx." Three-fourths or so of our space probes sent to Mars never make it there at all, mainly because of the vast distance, problems with radiation, mechanical failure, loss of communication, micrometeors, etc. Even so, the United States has a much better track record in this regard than the Russians, who have suffered fourteen failed attempts to reach the Red Planet.

Another issue is the length of the journey to Mars. Going to the moon with the Apollo program only took three days, but a one-way voyage to Mars will take upward of nine months, and a complete round-trip will take roughly two years. I once toured the NASA training center outside Cleveland, Ohio, where teams of scientists analyze the stresses of space travel. Astronauts suffer from muscle and bone atrophy caused by weightlessness when they spend extended periods in outer space. Our bodies are fine-tuned to live on a planet with the gravity of Earth. If the Earth were even a few percentage points larger or smaller, our bodies would have to be redesigned to survive on it. The longer we are in outer space, the more our bodies deteriorate. Russian astronaut Valeri Polyakov, after setting a world record for being in space for 437 days, could barely crawl out of his space capsule when he returned.

An interesting fact is that astronauts become several inches taller in outer space due to the expansion of their spinal columns. Once back on Earth, their height reverts back to normal. Astronauts may also lose 1 percent of their bone mass per month in space. To slow down this loss, they have to exercise at least two hours a day on a treadmill. Still, it can take astronauts a full year to rehabilitate after a six-month tour on the International Space Station—and sometimes, they never fully recover their bone mass. (A further consequence of weightlessness that wasn't taken seriously until recently is damage to the optic nerve. In the past, astronauts noted that their eyesight deteriorated after long missions in space. Detailed scans of their eyes show that their optic nerves are often inflamed, probably due to pressure from the fluid of the eye.)

In the future, our space capsules may have to spin so that the centrifugal force can generate artificial gravity. We experience this effect every time we go to a carnival and enter the spinning cylinder of a Rotor or Gravitron. The centrifugal force produces artificial gravity and pushes us back to the cylinder's wall. At present, a spinning spaceship would be too expensive to produce, and the concept is difficult to execute. The rotating cabin would have to be quite large, or else the centrifugal force would not be evenly distributed, and astronauts would get seasick and disoriented.

There is also the problem of radiation in space, especially from the solar wind and cosmic rays. We often forget that the Earth is blanketed by a thick atmosphere and covered with a magnetic field that helps to shield us. At sea level, our atmosphere absorbs most of the deadly radiation, but even on a normal plane ride across the United States, we receive an extra millirem of radiation per hour in the jet—the equivalent of a dental X-ray every time we take a cross-country flight. Astronauts traveling to Mars would have to pass through radiation belts that surround the Earth, which could expose them to heavy doses of radiation and increase their susceptibility to disease, premature aging, and cancer. Being on a two-year interplanetary trip, an astronaut would receive about two hundred times the radiation of a twin who stayed on the Earth. (However, this statistic should be placed in context. The astronaut's lifetime risk of developing cancer would rise from 21 percent to 24 percent. While not insignificant, this threat pales in comparison to the far-greater danger faced by an astronaut from a simple accident or mishap.)

Cosmic rays from outer space are sometimes so intense that astronauts can actually see tiny flashes of light as subatomic particles ionize the fluid in their eyeballs. I've interviewed several astronauts who have described these flashes, which look beautiful but can cause serious radiation damage to the eye.

And 2016 brought bad news concerning the effects of radiation on the brain. Scientists at the University of California, Irvine, exposed mice to large doses of radiation equivalent to the amount that would be absorbed during a two-year ride through deep space. They found evidence of irreversible brain damage. The mice showed behavioral problems and became agitated and dysfunctional. At the very least, these results confirm that astronauts must be adequately shielded in deep space.

In addition, astronauts have to worry about giant solar flares. In 1972, when Apollo 17 was being readied for a trip to the moon, a powerful solar flare hit the lunar surface. If the astronauts had been walking on the moon at the time, they might have been killed. Unlike cosmic rays, which are random, solar flares can be tracked from the Earth, so it is possible to warn astronauts several hours ahead of time. There have been incidents where the astronauts on the International Space Station were notified of approaching solar flares and ordered to move to better-protected sections of the Space Station.

Then, there are the micrometeors, which can tear the outer hull of a spacecraft. Close examination of the space shuttle reveals the impact of numerous micrometeorites on its tiled surface. The force of a micrometeor the size of a postage stamp traveling at forty thousand miles per hour would be enough to rip a hole in the rocket and cause rapid depressurization. It may be wise to separate space modules into different chambers, so that a punctured section can be rapidly sealed off from the others.

Psychological difficulties will present a different kind of obstacle. Being locked up in a tiny, cramped capsule with a small group of people for an extended period of time will be challenging. Even with a battery of psychological tests, we cannot definitively predict how—or whether—people will cooperate. Ultimately, your life may depend on someone who gets on your nerves.

GOING TO MARS

After months of intense speculation, in 2017 NASA and Boeing finally revealed the details of the plan to reach Mars. Bill Gerstenmaier, of NASA's Human Exploration and Operations Directorate, revealed a surprisingly ambitious timetable for the steps necessary to send our astronauts to the Red Planet.

First, after years of testing, the SLS/Orion rocket will be launched in 2019. It will be fully automatic, carrying no astronauts, but will orbit the moon. Four years later, after a fifty-year gap, astronauts will finally return to the moon. The mission will last three weeks, but it will just orbit around the moon, not land on the lunar surface. This is mainly to test the reliability of the SLS/Orion system rather than to explore the moon.

But there is an unexpected twist to NASA's new plan that surprised many analysts. The SLS/Orion system is actually a warm-up act. It will serve as the main link by which astronauts will leave the Earth and reach outer space, but an entirely new set of rockets will take us to Mars.

First, NASA envisions building the Deep Space Gateway, which resembles the International Space Station, except it is smaller and orbits the moon, not the Earth. Astronauts will live on the Deep Space Gateway, which will act as a refueling and resupply station for missions to Mars and the asteroids. It will be the basis for a permanent human presence in space. Construction of this lunar space station will begin in 2023 and it will be operational by 2026. Four SLS missions will be required to build it.

But the main act is the actual rocket that will send astronauts to Mars. It is an entirely new system, called the Deep Space Transport, which will be constructed mainly in outer space. In 2029, the Deep Space Transport will have its first major test, circling around the moon for three hundred to four hundred days. This will provide valuable information about long-term missions in space. Finally, after rigorous testing, the Deep Space Transport will send our astronauts to orbit Mars by 2033.

NASA's program has been praised by many experts because it is methodical, with a step-by-step plan to build an elaborate infrastructure on the moon.

However, NASA's plan stands in sharp contrast to Musk's vision. NASA's plan is carefully fleshed out and involves the creation of a permanent infrastructure in lunar orbit, but it is slow, perhaps taking a decade longer than Musk's plan. SpaceX bypasses the lunar space station entirely and blasts directly to Mars, perhaps as early as 2022. One drawback, however, of Musk's plan is that the Dragon space capsule is considerably smaller than the Deep Space Transport. Time will tell which approach or combination of approaches is better.

FIRST TRIP TO MARS

Because more details are being revealed concerning the first Mars mission, it is now possible to speculate on the steps necessary to reach the Red Planet. Let us trace how NASA's plan may unfold over the next few decades.

The first people on the historic mission to Mars are probably alive today, perhaps learning about astronomy in high school. They will be among the hundreds of people who are expected to volunteer for the first mission to another planet.

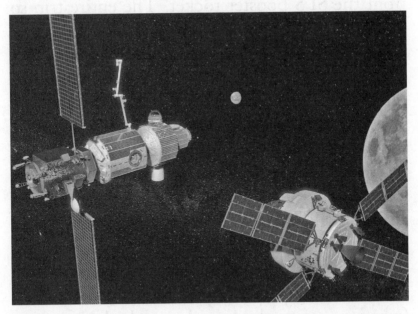

NASA's Deep Space Gateway will orbit the moon and serve as a fuel and supply station for missions to Mars and beyond.

After rigorous training, perhaps four candidates will be carefully chosen for their skills and experience, probably including a seasoned pilot, an engineer, a scientist, and a doctor.

Sometime around 2033, after a series of anxious interviews with the press, they will finally climb aboard the Orion space capsule. Although the Orion has 50 percent more room than the original Apollo capsule, things will still be cramped inside, but it doesn't matter, since the trip to the moon will last only three days. When the spaceship finally blasts off, they will feel vibrations from the intense burning of rocket fuel from the SLS booster rocket. The entire trip so far looks and feels very similar to the original Apollo mission.

But here, the similarity ends. From this point, NASA envisions a radical departure from the past. As they enter lunar orbit, the astronauts will see the Deep Space Gateway, the world's first space station orbiting the moon. The astronauts will dock with the Gateway and rest a bit.

They will then transfer to the Deep Space Transport, which looks like no other spacecraft in history. The spaceship and crew's quarters resemble a long pencil, with an eraser at one end (which contains the capsule in which the astronauts will live and work). Along the pencil, there are series of gigantic arrays of unusually long, narrow solar panels, so, from a distance, the rocket begins to resemble a sailboat. While the Orion capsule weighs about twenty-five tons, the Transport weighs forty-one.

The Deep Space Transport will be their home for the next two years. This capsule is much bigger than the Orion and will give astronauts enough room to stretch a bit. This is important, since they have to exercise daily to prevent muscle and bone loss, which could cripple them when they reach Mars.

Once on board the Deep Space Transport, they will turn on the rocket's engines. But instead of being jolted by a powerful thrust and watching gigantic flames shoot from the back of the rocket, the ion engines will accelerate smoothly, gradually building up speed. Staring outside their windows, the astronauts will only see the gentle luminous glow of hot ions being steadily emitted from the ship's engines.

The Deep Space Transport uses a new type of propulsion system to send astronauts through space, called solar electric propulsion. The huge solar panels capture sunlight and convert it to electricity. This is used to strip away the electrons from a gas (like xenon), creating ions. An electric field then shoots these charged ions out one end of the engine, creating thrust. Unlike chemical engines, which can only fire for a few minutes, ion engines can slowly accelerate for months or even years.

Then begins the long, boring trip to Mars itself, which will take about nine months. The main problem facing the astronauts is boredom, so they will have to constantly exercise, play games to keep alert, do calculations, talk to their loved ones, surf the web, etc. Other than routine course corrections, there is not much else to do during the actual voyage. Occasionally, however, they might be required to do some spacewalks in order to make minor repairs or replace worn parts. As the journey progresses, however, the time it takes to send radio messages to Earth gradually increases, eventually reaching about twenty-four minutes. This may prove a bit frustrating for the astronauts, who are used to instantaneous communication.

As they gaze out their windows, they will gradually see the Red Planet come into focus, looming in front of them. Activity aboard the spaceship will rapidly quicken as the astronauts begin to make preparations. At this point, they will fire their rockets to slow their spacecraft down so they can gently enter into orbit around Mars.

From space, they will see an entirely different panorama than seen on the Earth. Instead of blue oceans, green tree-covered mountains, and the lights of cities, they will see a barren, desolate landscape, full of red deserts, majestic mountains, gigantic canyons that are much larger than the ones on Earth, and huge dust storms, some of which can engulf the entire planet.

Once in orbit, they will enter the Mars capsule and separate from the main spacecraft, which will continue to orbit the planet. As their capsule enters into the Martian atmosphere, the temperature will rise dramatically, but the heat shield will absorb the intense heat generated by air friction. Eventually, the heat shield will be ejected, and the capsule will then fire its retrorockets and slowly descend onto the surface of Mars.

Once they exit the capsule and walk on the surface of Mars, they will be pioneers opening up a new chapter in the history of the human race, taking a historic step toward realizing the goal of making humanity a multiplanet species.

They will spend several months on the Red Planet before the Earth is in the right alignment for the return trip. This will give them time to scout the terrain, do experiments, such as looking for traces of water and microbial life, and set up solar panels for power. One possible objective might be to drill for ice in the permafrost, since underground ice may one day become a vital source of drinking water, as well as oxygen for breathing and hydrogen for fuel.

After their mission is complete, they will go back into their space capsule and then blast off. (Because of Mars's weak gravity, the capsule requires much less fuel than it would to leave the Earth.) They will dock with the main ship in orbit, and then the astronauts will prepare for the nine-month journey back to the Earth.

Upon their return, they will splash down somewhere in the ocean. Once back on terra firma, they will be celebrated as heroes who took the first step toward establishing a new branch of humanity.

As you can see, we will face many challenges on the road to the Red Planet. But with the public's enthusiasm, and with the commitment of NASA and the private sector, it is likely that we will achieve a manned mission to Mars in the next decade or two. This will open up the next challenge: to transform Mars into a new home.

> I think that when humans get around to exploring and building cities and towns on Mars, it will be viewed as one of the great times of humanity, a time when people set foot on another world and had the freedom to make their own world.
> —ROBERT ZUBRIN

5 MARS: THE GARDEN PLANET

In the 2015 movie **The Martian,** the astronaut played by Matt Damon faces the ultimate challenge: to survive alone on a frozen, desolate, airless planet. Accidentally left behind by his fellow crewmates, he has only enough supplies to last a few days. He must summon all his courage and know-how to last until a rescue mission can reach him.

The movie was realistic enough to give the public a taste of the difficulties Martian colonists would encounter. For one, there are the fierce dust storms, which engulf the planet with a fine red dust that resembles talcum powder and almost tipped over the spacecraft in the movie. The atmosphere is almost entirely made of carbon dioxide, and the atmospheric pressure is only 1 percent that of the Earth, so an astronaut would suffocate within a few minutes if exposed to the thin Martian air, and his blood would begin to boil. To produce enough oxygen to breathe, Matt Damon has to create a chemical reaction in his pressurized space station.

And since he is rapidly running out of food, he has to grow his own plants in an artificial garden. To fertilize the crops, he has to use his own waste.

Bit by bit, the astronaut in **The Martian** takes the excruciating steps necessary to create an ecosystem on Mars that is capable of sustaining him. The movie helped to capture the imagination of a new generation. But the fascination with Mars actually has a long and interesting history that stretches back to the nineteenth century.

In 1877, Italian astronomer Giovanni Schiaparelli noticed strange linear markings on Mars that seemed to be formed by natural processes. He called the markings "**canali**," or channels. However, when the Italian was translated into

English, the i was dropped and the term became "canals," which has an entirely different meaning: they are artificial, not natural. A simple mistranslation gave way to an avalanche of speculation and fantasy, sparking the "man from Mars" myth. The wealthy, eccentric astronomer Percival Lowell began to theorize that Mars was a dying planet and that the Martians had dug the canals in a desperate attempt to transport water from the polar ice caps to irrigate their scorched fields. Lowell would dedicate his life to proving his conjecture, using his considerable private fortune to build a state-of-the-art observatory in Flagstaff in the Arizona desert. (He never did prove the existence of these canals, and years later, space probes would show that the canals were an optical illusion. But the Lowell Observatory scored successes in other areas, contributing to the discovery of Pluto and providing the first indication that the universe was expanding.)

In 1897 H. G. Wells wrote **The War of the Worlds.** The Martians in the novel plan to annihilate humanity and "terraform" the Earth so that its climate becomes like that of Mars. The book gave rise to a new literary genre—you could call it the "Mars attacks" genre—and the idle, esoteric discussions of professional astronomers suddenly became a matter of survival for the human race.

On the day before Halloween in 1938, Orson

Welles took excerpts from the novel to create a series of short, dramatic, realistic radio broadcasts. The program was presented as if the Earth was actually being invaded by hostile Martians. Some people began to panic, hearing updates on the invasion—how the armed forces had been overwhelmed by death rays, and how the Martians were converging on New York City in giant tripods. Rumors from terrified listeners spread rapidly across the country. In the aftermath of this chaos, the major media vowed never again to broadcast a hoax as if it were real. This ban continues today.

Many people were caught up in Martian hysteria. The young Carl Sagan was enthralled by novels about Mars, such as the John Carter of Mars series. In 1912, Edgar Rice Burroughs, famous for his Tarzan novels, dabbled in science fiction by writing about an American soldier during the Civil War who is transported to Mars. Burroughs speculated that John Carter would become a superman because of the low gravity on Mars relative to Earth. He would be able to jump incredible distances and outfight the alien Tharks to save the beautiful Dejah Thoris. Cultural historians believe that this explanation for the superpowers of John Carter formed the basis of the Superman story. The 1938 issue of **Action Comics** in which Superman first appears attributes his superpowers to the weak gravity of the Earth compared to his native Krypton.

LIVING ON MARS

Taking up residence on Mars may sound romantic in science fiction, but the realities are quite daunting. One strategy for prospering on the planet is to take advantage of what is available, such as ice. Since Mars is frozen solid, all you would have to do is dig a few feet until you hit the permafrost. Then you could excavate the ice, melt it, and purify it for drinking water, or extract oxygen for breathing and hydrogen for heating and rocket fuel. For protection against radiation and dust storms, colonists might have to dig into the rock to build an underground shelter. (Because the atmosphere of Mars is so thin and its magnetic field is so weak, radiation from space is not absorbed or deflected as it is on Earth, so this is a real problem.) Or it could be advantageous to set up the first Martian base in a gigantic lava tube near a volcano, as we discussed doing on the moon. Given the prevalence of volcanoes on Mars, it is likely such tubes would be plentiful.

A day on Mars is roughly the same duration as a day on Earth. The tilt of Mars with respect to the sun is also the same as Earth's. But settlers would have to get used to the gravity on Mars, which is only 40 percent of the gravity on Earth, and, as on the moon, they would have to exercise vigorously to avoid muscle and bone loss. They would also need to contend with the brutally cold weather and would be in a constant struggle to avoid freezing to death. The temperature on Mars rarely exceeds the freezing point of water, and after the sun goes down, it can plunge to as low as -127 degrees Celsius or -197 degrees Fahrenheit, so any power failure or blackout could prove life threatening.

Even if we can send the first manned mission to Mars by 2030, because of these obstacles it may take until 2050 or beyond to compile sufficient equipment and supplies to create a permanent outpost on the planet.

MARTIAN SPORTS

Because of the vital importance of exercise to prevent muscle deterioration, astronauts on Mars will necessarily have to engage in vigorous sports, where they will find, much to their delight, that they have superhuman abilities.

But this also means that sports arenas would have to be completely redesigned. Because the gravity on Mars is a little bit more than one-third the gravity on Earth, a person can in principle jump three times higher on Mars. A person would also be able to throw a ball three times farther on Mars, so basketball courts, baseball diamonds, and football fields would have to be enlarged.

Furthermore, the atmospheric pressure on Mars is about 1 percent that of Earth, meaning that the aerodynamics of baseballs and footballs are drastically modified. The main complication is the precise control of the ball. On Earth, athletes are paid millions of dollars because of their uncanny ability to control the motion of a ball, which takes years of practice. This skill has to do with their ability to manipulate the ball's spin.

When a ball moves through the air, it creates turbulence in its wake, small eddy currents that cause the ball to swerve slightly and slow down. (For a baseball, these eddy currents are created by the stitching on the ball, which determines its spin. On a golf ball, it is caused by the dimples on its surface. For soccer balls, it is due to the juncture between the plates on its surface.)

Football players throw the ball so that it spirals rapidly in the air. Spinning reduces the eddy currents on the ball's surface, so it can more accurately slice through the air and travel much farther without tumbling. Also, because it is spinning rapidly, it is like a small gyroscope and hence points steadily in one direction, which keeps the football moving in the correct path and makes it easier to catch.

Using the physics of airflow, it is also possible to show that many of the myths concerning throwing a baseball are true. For generations, baseball pitchers have claimed that they can throw knuckleballs and curveballs, which allows them to control the ball's trajectory, seemingly in violation of common sense.

Time-lapse videos show that this is correct. If a baseball is thrown so that it has minimal spin (as in a knuckleball), turbulence is maximized and the ball's path becomes erratic. If a baseball is spinning rapidly, then the air pressure on one side of the baseball can be greater than the pressure on the other side (via something called Bernoulli's principle) and hence the ball will swerve a certain way.

All this means that, for world-class athletes from Earth, the reduced air pressure on Mars may cause them to lose their ability to control the ball, so that an entirely new crop of Martian athletes may rise in their place. Mastery of a sport on Earth may mean little when applied to Mars.

If we draw up a list of the sports that are found in the Olympics, we see that, without exception, each and every one would have to be modified to take into account the reduced gravity and air pressure on Mars. In fact, a new Martian Olympics may emerge, including radical sports that are not physically possible on Earth and don't even exist yet.

The conditions on Mars may also increase the artistry and elegance of other sports. A figure skater, for example, can only spin about four times in the air on the Earth. No skater has ever performed a quintuple jump. This is because the height of the jump is determined by the velocity at takeoff and the strength of gravity. On Mars, figure skaters will be able to soar three times higher in the air and execute breathtaking jumps and spins because of the reduced gravity and air pressure. Gymnasts on Earth perform marvelous twists and turns in the air because their muscle strength exceeds the weight of their body. But on Mars, their strength would vastly exceed their reduced body weight, allowing them to perform twists and turns in the air that have never been seen before.

TOURISTS ON MARS

Once our astronauts have mastered the fundamental life-and-death challenges of surviving on Mars, they can savor some of the aesthetically pleasing rewards of the Red Planet.

Because of the planet's weak gravity, thin atmosphere, and lack of liquid water, Martian mountains can grow to truly majestic proportions compared to Earth-bound ones. Mars's Olympus Mons is the largest known volcano in the solar system. It is about 2.5 times taller than Mt. Everest and so wide that, if placed on North America, it would extend from New York City to Montreal, Canada. The low gravity field also means that mountain climbers would not be burdened by heavy backpacks and would be able to perform prodigious feats of endurance, like astronauts on the moon.

Adjacent to Olympus Mons are three smaller volcanoes in a straight line. The presence and position of these smaller volcanoes are indicative of ancient tectonic activity on Mars. The Hawaiian Islands here on Earth provide a useful analogy. There is a stationary pool of magma under the Pacific Ocean, and as the tectonic plate moves over this magma pool, the pressure from the magma periodically pushes upward through the crust, creating the latest island in the Hawaiian chain. But tectonic activity seems to have ended on Mars long ago, providing evidence that the core of the planet has cooled down.

The largest Martian canyon, Mariner Valley, which is probably the largest canyon in the solar system, is so vast that, if placed on North America, it would extend from New York City to Los Angeles. Hikers who have marveled at the Grand Canyon would be astounded by this extraterrestrial canyon network. But unlike the Grand Canyon, Mariner Valley does not have a river at the bottom. The latest theory is that the more-than-three-thousand-mile canyon is the juncture of two ancient tectonic plates, like the San Andreas Fault.

A prime tourist attraction will be the Red Planet's two giant polar ice caps, which feature two kinds of ice and differ in composition from those on the Earth. One type of ice cap is made of frozen water. These are a permanent fixture on the landscape and remain roughly the same for much of the Martian year. The other variety consists of dry ice, or frozen carbon dioxide, and they expand or contract depending on the season. In the summer, the dry ice vaporizes and disappears, leaving only the ice caps composed of water. As a result, the appearance of the polar ice caps will vary during the course of the year.

Whereas the Earth's surface is continually changing, Mars's basic topography has not altered much in a few billion years. As a result, Mars has features that have no counterpart on Earth, including remnants of thousands of giant meteor craters that were formed long ago. The Earth once had giant meteor craters as well, but water erosion erased many of them. Furthermore, most of the surface of the Earth is recycled every few hundred million years due to tectonic activity, so ancient craters have all been transformed into new terrain. Looking at Mars, however, is gazing at a landscape frozen in time.

In many ways, we actually know more about the surface of Mars than the surface of the Earth. About three-quarters of the Earth is covered by the oceans, while Mars has no oceans. Our Mars orbiters have been able to photograph every square meter of its surface and give us detailed maps of its terrain. The combination of ice, snow, dust, and sand dunes on Mars creates all sorts of novel geologic formations that are not seen on Earth. Walking across the Martian terrain would be a hiker's dream.

One apparent hindrance to making Mars a tourist destination might be the monster dust devils, which are quite common and can be seen crisscrossing the deserts almost daily. They can be taller than Mt. Everest, dwarfing those on Earth, which only rise a few hundred feet into the air. There are also ferocious, planet-sized dust storms that can envelop all of Mars in a blanket of sand for weeks. But they would not do much damage, thanks to the planet's low atmospheric pressure. Hundred-mile-an-hour winds would only feel like a ten-mile-an-hour breeze to an astronaut. They may be a nuisance, blowing fine particles into our space suits, machinery, and vehicles and causing malfunctions and breakdowns, but they are not going to topple buildings and structures.

Because the air is so thin, airplanes would need a much larger wingspan to fly on Mars than on Earth. A solar-powered aircraft would require tremendous surface area and might be too expensive to deploy for recreational purposes. We probably will not see tourists flying through Martian canyons like they do over the Grand Canyon. But balloons and blimps could be a viable means of transportation, in spite of the low temperature and low atmospheric pressure. They could explore the Martian terrain at much closer distances than orbiters, yet still cover vast areas of the surface. One day, fleets of balloons and blimps may be a regular sight over the geologic wonders there.

MARS: A GARDEN OF EDEN

To maintain a lasting presence on the Red Planet, we must find a way to create a Garden of Eden on its inhospitable landscape.

Robert Zubrin, an aerospace engineer who has worked with Martin Marietta and Lockheed Martin, is also founder of the Mars Society and for years has been one of the most vocal proponents of colonizing the Red Planet. His aim is to convince the public to fund a manned mission. Once, he was a lone voice, pleading with anyone who would listen, but now, companies and governments are seeking his advice.

I have interviewed him on several occasions, and each time his enthusiasm, energy, and dedication to his mission shined through. When I asked him what sparked this fascination with space, he told me that it all started with reading science fiction as a child. He also was mesmerized when, as early as 1952, von Braun showed how a mission of ten spaceships, assembled in orbit, could take a crew of seventy astronauts to Mars.

I asked Dr. Zubrin how his fascination with science fiction translated into a lifelong quest to reach Mars. "Actually, it was Sputnik," he said. "To the adult world, it was terrifying, but to me it was exhilarating." He was captivated by the 1957 launch of the world's first artificial satellite because it meant that the novels he was reading might come true. Science fiction, he firmly believed, would one day become science fact.

Dr. Zubrin was part of the generation that saw the United States start from scratch to become the greatest space-faring nation on the planet. Then people began to be consumed by the Vietnam War and internal strife, and walking on the moon seemed increasingly distant and unimportant. Budgets were slashed. Programs were canceled. Although the public mood turned against the space program, Dr. Zubrin maintained his conviction that Mars should be the next milestone on our agenda. In 1989, President George H. W.

Bush briefly excited the public imagination by announcing plans to reach Mars by 2020—until the following year, when studies showed that the price tag for the project would be about $450 billion. Americans got sticker shock, and the Mars mission was shelved once again.

Zubrin spent years wandering in the wilderness, trying to drum up support for his ambitious agenda. Acknowledging that the public would not support any scheme that was over budget, Zubrin proposed a number of novel but realistic approaches to colonizing the Red Planet. Before he came along, most people did not seriously consider the problem of financing future space missions.

In his 1990 proposal, called Mars Direct, Zubrin reduced costs by splitting the mission into two parts. Initially, an unmanned rocket called the Earth Return Vehicle is sent to Mars. It carries a small amount of hydrogen—only 8 tons' worth—but combines it with the unlimited supply of carbon dioxide that occurs naturally in the Martian atmosphere. This chemical reaction produces up to 112 tons of methane and oxygen and provides enough rocket fuel for the subsequent return voyage. Once the fuel has been generated, astronauts take off in a second vehicle called the Mars Habitat Unit, which contains only enough fuel for a one-way trip to Mars. After the astronauts land, they conduct scientific experiments. Then they leave the Mars Habitat Unit and transfer into the Earth Return Vehicle from the original mission, which is loaded with the newly created rocket fuel. This ship would then bring them back to Earth.

Critics are sometimes horrified to hear that Zubrin advocated giving travelers a one-way ticket to Mars, as if expecting them to die on the Red Planet. He is careful to explain that the fuel for the return trip can be manufactured on Mars. But he adds, "Life is a one-way trip, and one way to spend it is by going to Mars and starting a new branch of human civilization there." He believes that five hundred years from now, historians may not remember all the petty wars and conflicts of the twenty-first century, but humanity will celebrate the founding of its new community on Mars.

NASA has since adopted aspects of the Mars Direct strategy, which changed the philosophy of the Mars program to prioritize cost, efficiency, and living off the land. Zubrin's Mars Society has also constructed a prototype of an actual Mars base. They chose Utah as the site for their Mars Desert Research Station (MDRS) because the environment came closest to simulating the conditions on the Red Planet: cold, deserted, barren, and lacking in vegetation and animals. The core of the MDRS is its habitat, a two-story cylindrical building that can house seven crew members. There is also a large observatory for stargazing. The MDRS takes volunteers from the public, who commit to a two- to three-week stay at the station. The volunteers are trained to behave as actual astronauts with certain obligations and duties, such as conducting science experiments, performing maintenance, and making observations. The organizers of the MDRS try to make the experience as realistic as possible and use these sessions as a way to test the psychological dimension of being isolated on Mars for extended periods with relative strangers. More than one thousand people have passed through the program since it began in 2001.

The lure of Mars is so strong that it has attracted several ventures of questionable value. The MDRS should not be confused with the Mars One program, which advertises a dubious one-way trip to Mars for those who pass a sequence of tests. Though hundreds have applied, the program has no concrete means of getting to Mars. It claims that it will pay for its rocket by soliciting donations and producing a movie about its mission. Skeptics charge that the leaders of the Mars One program are better at conning the press than attracting genuine scientific expertise.

Another outlandish attempt to form an isolated colony like one we would create on Mars was a project called Biosphere 2, bankrolled by $150 million from the Bass family fortune. A three-acre domed complex made of glass and steel was erected in the Arizona desert. It could house eight humans and three thousand plant and animal species and was meant to serve as a sealed habitat to test whether humans could survive in a controlled, isolated environment that resembles what we might one day create on another planet. From its start in 1991, the experiment was plagued with a series of mishaps, disputes, scandals, and malfunctions that generated more headlines than real science. Fortunately, the facilities were taken over by the University of Arizona in 2011, and since then they have become a valid research center.

TO TERRAFORM MARS

Based on his experience with MDRS and other efforts, Dr. Zubrin predicts that the colonization of Mars will proceed in a predictable sequence. In his view, the first priority is to establish a base for around twenty to fifty astronauts on the surface of Mars. Some would stay for only a few months. Others would become lifers and make the base their permanent home. Over time, the people on Mars would start to see themselves less as astronauts and more as settlers.

Most supplies would initially have to come

from Earth, but in the second phase, the population would rise to a few thousand people, and they would become capable of exploiting the raw materials of the planet. The red color of the sands on Mars is due to the presence of iron oxide, or rust, so settlers would be able to make iron and steel for construction. Electricity can be generated from large solar parks harvesting energy from the sun. The carbon dioxide in the atmosphere could be used to cultivate plants. The Mars settlement would gradually become self-sufficient and sustainable.

The next step is the most difficult of all. Ultimately, the colony will have to find a way to slowly heat the atmosphere so that liquid water can flow on the Red Planet for the first time in three billion years. This would make agriculture and, eventually, cities possible. At that point, we would enter the third stage, and a new civilization could flourish on Mars.

Rough calculations suggest that it may be prohibitively expensive at present to terraform Mars and that it would take centuries to complete the process. However, what is intriguing and promising about the planet is the geographic evidence that liquid water was once abundant on the surface, etching riverbeds, riverbanks, and even the outline of an ancient ocean the size of the United States. Billions of years ago, Mars cooled down before the Earth did and had a tropical climate when the Earth was still molten. This combination of mild weather and large bodies of water has led some scientists to speculate that DNA originated on Mars. In this scenario a giant meteor impact blasted tremendous amounts of debris into outer space—some of it later landing on Earth and seeding it with Martian DNA. If this theory is correct, then all you have to do to see a Martian is look in the mirror.

Zubrin points out that terraforming is not a new or strange process. After all, the DNA molecule is continually terraforming the Earth. Life has reshaped every aspect of the Earth's ecology, from the composition of the atmosphere, to the Earth's topography, to the makeup of the oceans. So we will simply be following nature's own script when we begin to terraform Mars.

JUMP-STARTING THE WARMING OF MARS

To initiate the process of terraforming, we might inject methane and water vapor into the atmosphere to induce an artificial greenhouse effect. These greenhouse gases would capture sunlight and steadily raise the temperature of the ice caps. As the ice caps melt, they would release trapped water vapor and carbon dioxide.

We might also send satellites into orbit around Mars to direct concentrated sunlight onto the ice caps. The satellites could be synchronized to hover over a fixed point in the sky and direct energy to the polar regions. On Earth, we angle our satellite TV dishes toward a similar geo-stationary satellite about twenty-two thousand miles away that appears to be fixed in the sky because it makes a complete revolution around the Earth every twenty-four hours. (Geostation-ary satellites are in orbit above the equator. This means that the energy from these satellites will either hit the poles at an angle, or the energy will have to be beamed down to the equator and then transported to the poles. Unfortunately, either alternative involves some loss of energy.)

In this scheme, these Martian solar satellites would unfurl gigantic sheets, many miles across, containing a vast array of mirrors or solar panels. The sunlight could either be focused and then aimed toward the ice caps, or the energy could be converted using solar cells and then sent down as microwaves. This is one of the most efficient, al-beit costly, approaches to terraforming, because it is safe, nonpolluting, and ensures minimal damage to the surface of Mars.

There have been other proposed strategies. We could consider mining methane-rich Titan, one of the moons of Saturn, and bringing the methane to Mars. The methane could contribute to the desired greenhouse effect—methane, for reference, is over twenty times more effective at trapping heat than carbon dioxide. Another possible method is to make use of nearby comets or asteroids. As we've discussed, comets are largely composed of ice, and asteroids are known to contain ammonia, a greenhouse gas. If they happen to pass Mars, they can be deflected slightly so that they orbit the planet. Then they can be further redirected until they execute a very slow death spiral toward Mars. As they enter the Martian atmosphere, friction heats them up until they disintegrate, releasing water vapor or ammonia. This trajectory would be a magnificent sight from the surface of Mars. In some sense, NASA's Asteroid Redirect Mission (ARM) can be thought of as a practice run for such an undertaking. The ARM, you recall, is a future NASA mission to either retrieve rock samples from or gently alter the trajectory of a comet or asteroid. Of course, this technology has to be fine-tuned or we risk deflecting a giant asteroid onto the surface of Mars and wreaking havoc on a colony.

A more unorthodox idea, suggested by Elon Musk, is to melt the ice caps by detonating hydrogen bombs high above them. This method is currently possible with off-the-shelf technology. In principle, hydrogen bombs, although highly

protected, are relatively inexpensive to manufacture, and we certainly have the technology to
drop scores of them onto the ice caps with existing
rockets. However, no one knows how stable the
ice caps are or what the long-term effects of this
procedure might be, and many scientists are un-
easy about the risk of unintended consequences.

It is estimated that, if the ice caps of Mars were
completely melted, there would be enough liquid
water to fill a planetary ocean fifteen to thirty
feet deep.

REACHING THE TIPPING POINT

These proposals all endeavor to bring the Martian atmosphere to a tipping point where the
warming would become self-sustaining. Rais-
ing the temperature by six degrees Celsius would
be sufficient to instigate the melting process.
The greenhouse gases emitted from the ice caps
would heat the atmosphere. The carbon dioxide
absorbed into the desert aeons ago would also be
released and contribute to planetary warming,
causing further melting. Thus, the heating of
Mars would continue without further intervention from the outside. The warmer the planet,
the more water vapor and greenhouse gases are
released, which in turn warms the planet even
more. This process could carry on almost indefi-

nitely and would increase Mars's atmospheric pressure.

Once liquid water starts to flow within the ancient riverbeds of Mars, settlers could begin large-scale agriculture. Plants love carbon dioxide, so the first outdoor crops might be raised, and their waste products could be used to generate a layer of topsoil. Another positive feedback loop would be initiated: more crops would produce more soil, which could be used to nurture additional crops. The native soil of Mars also contains valuable nutrients such as magnesium, sodium, potassium, and chlorine that would help plants succeed. As plants begin to proliferate, they will also generate oxygen, an essential ingredient for terraforming Mars.

Scientists have created greenhouses that simulate the harsh conditions on Mars to see if plants and bacteria can survive there. In 2014, NASA's Institute for Advanced Concepts partnered with Techshot to construct biodomes with controlled environments in which to grow oxygen-producing cyanobacteria and algae. Preliminary tests indicate that certain life-forms can indeed flourish there. In 2012, scientists at the Mars Simulation Laboratory, maintained by the German Aerospace Center, found that lichen, which is similar to moss, could survive there for at least a month. In 2015, scientists at the University of Arkansas showed that four species of methanogens, micro-organisms that produce methane, can survive in a habitat resembling the Martian ecology.

Even more ambitious is NASA's Mars Ecopoiesis Test Bed, a project that aims to send hardy bacteria and plants, such as extremophile photosynthetic algae and cyanobacteria, to Mars aboard a rover. These life-forms would be placed in canisters that could be drilled down into the Martian soil. Water would be added to the canisters, and then instruments would look for the presence of oxygen, which would indicate active photosynthesis. If this experiment is successful, Mars may one day be covered with farms of this kind to generate oxygen and food.

By the beginning of the twenty-second century, the technologies of the fourth wave—nanotech, biotech, and AI—should be mature enough to have a profound impact on the terraforming of Mars.

Some biologists have posited that genetic engineering may result in a new species of alga that is designed to exist on Mars, perhaps in the particular chemical mix of its soil or in newly formed lakes. This alga would thrive in the cold, thin, carbon-dioxide-rich atmosphere and release copious quantities of oxygen as a waste product. It would be edible and could be bioengineered to mimic flavors found on Earth. In addition, it would be engineered to produce an ideal fertilizer.

In the movie **Star Trek II: The Wrath of Khan,**

a fantastic new technology called the Genesis Device was introduced. It was capable of terraforming dead planets into lush, livable worlds almost instantly. It would explode like a bomb and release a spray of highly bioengineered DNA. As this super DNA spreads to all corners of the planet, the cells would take root and dense jungles would form until the whole planet was terraformed within a matter of days.

In 2016, Claudius Gros, a professor at Goethe University in Frankfurt, Germany, published a paper in the journal **Astrophysics and Space Science** detailing what a real-life Genesis Device might look like. He predicts that a primitive version will be possible in fifty to one hundred years. First, scientists on Earth would have to carefully analyze the ecology of the lifeless planet. The temperature, soil chemistry, and atmosphere would determine which types of DNA should be introduced. Then, fleets of robotic drones would be sent to the planet to deposit millions of nano-sized descent capsules carrying an array of DNA. When these capsules release their contents, the DNA, engineered precisely to thrive in the planet's environmental conditions, would latch onto the soil and begin to germinate. The contents of these capsules are designed to reproduce by creating seeds and spores on the barren planet and use the minerals found there to create vegetation.

Dr. Gros believes that life on the newly seeded planet would have to develop the old-fashioned way, by evolution. He warns that "global-scale ecological disasters" might occur if we try to rush this process, especially if one type of life-form ends up proliferating so rapidly that it pushes out the others.

WILL TERRAFORMING LAST?

If we succeed in terraforming Mars, what is to prevent it from reverting back to its original barren state? Investigating this issue brings us back to a critical question that has been nagging at astronomers and geologists for decades: Why did Venus, Earth, and Mars evolve so differently?

When the solar system formed, the three planets were similar in many ways. They had volcanic activity, which released large quantities of carbon dioxide, water vapor, and other gases into their atmospheres. (This is why, even today, the atmospheres of Venus and Mars consist almost exclusively of carbon dioxide.) The water vapor condensed into clouds, and the rain helped to carve out the rivers and lakes. If they had been closer to the sun, their oceans would have boiled away; and if they were farther out, their oceans would have frozen. But all three were within or very close to the "Goldilocks zone," the band around a star that allows water to remain in liquid form. Liquid water is the "universal solvent" out of which the first organic chemicals materialized.

Venus and the Earth are almost identical in size. They are celestial twins, and by rights, they should have followed the same evolutionary history. Science fiction writers once envisioned Venus as a verdant world that would make a perfect vacation spot for weary astronauts. In the 1930s, Edgar Rice Burroughs introduced another interplanetary swashbuckler, Carson Napier, in **Pirates of Venus,** which described the planet as a jungle-like wonderland, full of adventure and danger. But today, scientists realize that Venus and Mars do not resemble the Earth at all. Something happened billions of years ago that sent these three planets on very distinct paths.

In 1961, when the romantic notion of a Venusian utopia still dominated the public imagination, Carl Sagan made the controversial conjecture that Venus suffered from a runaway greenhouse effect and was devilishly hot. His novel and disturbing theory was that carbon dioxide acts as a one-way street for sunlight. Light can readily enter through the carbon dioxide in Venus's atmosphere because the gas is transparent. But once the light bounces off the ground, it turns into heat or infrared radiation, which cannot easily escape the atmosphere. The radiation becomes trapped, in a process similar to the way a greenhouse captures sunlight during winter or the way cars heat up in the summer sun. This process happens on the Earth, but it is vastly accelerated on Venus because it is much closer to the sun, and a runaway greenhouse effect was the result.

Sagan was proven correct the next year when the Mariner 2 probe flew past Venus and revealed something truly shocking: the temperature was a blistering nine hundred degrees Fahrenheit, hot enough to melt tin, lead, and zinc. Instead of being a tropical paradise, Venus was a hellhole resembling a blast furnace. Subsequent space shots confirmed the bad news. And there was no relief when it rained, because the rains consist of caustic sulfuric acid. Considering that Venus is linked to the Greek goddess of love and beauty, it is ironic that this sulfuric acid, which is highly reflective, is the reason why Venus shines so brightly in the night sky.

In addition, the atmospheric pressure of Venus was found to be almost one hundred times that of the Earth. The greenhouse effect helps to explain why. Most of the carbon dioxide on the Earth is recycled, dissolving in the oceans and in rocks. But on Venus, the temperature became so high that the oceans boiled off. And instead of dissolving in rocks, the gas was baked out of them. The more carbon dioxide outgassed from the rocks, the hotter the planet got, setting off a feedback loop.

Due to the planet's high atmospheric pressure, being on the surface of Venus is equivalent to being three thousand feet below the surface of Earth's oceans. You would be crushed like an eggshell. But if you could find a way to overcome this and the searing temperatures, you would still be confronted with a scene from Dante's **Inferno.** The air is so dense that, when walking on the surface, you would have the sensation of walking through molasses, and the ground under your feet would feel soft and squishy because it is made of molten metal. The acid rains would eat through the tiniest tear in your space suit, and one false move and you might sink into a vat of molten magma.

Given these constraints, terraforming Venus seems out of the question.

WHAT HAPPENED TO MARS'S OCEAN?

If our twin, Venus, turned out differently because it was closer to the sun, how do we explain the evolution of Mars?

The key is that Mars is not only farther from the sun, but it is also much smaller and therefore cooled off faster than the Earth. Its core is no longer molten. Planetary magnetic fields are generated by the motion of metal within a liquid core, creating electrical currents. Since the core of Mars is made of solid rock, it cannot create an appreciable magnetic field. In addition, it is believed that heavy meteor bombardment three or so billion years ago triggered so much chaos that the original magnetic field was disrupted. This may explain why Mars lost its atmosphere and water. Without a magnetic field to protect it against harmful solar rays and flares, the atmosphere was gradually blown into outer space by the solar wind. As the atmospheric pressure dropped, the oceans boiled away.

Another process accelerated the loss of its atmosphere. Much of the original carbon dioxide on Mars dissolved into the oceans and turned into carbon compounds, which subsequently were deposited on the ocean floor. Tectonic activity on the Earth periodically recycles the continents and enables carbon dioxide to rise to the surface again. But because the core of Mars is probably solid, it has no significant tectonic activity, and its carbon dioxide was locked into the ground permanently. As carbon dioxide levels began to drop, a reverse greenhouse effect took place and the planet went into a deep freeze.

The dramatic contrasts between Mars and

Venus can help us appreciate the Earth's geologic history. The core of the Earth could have cooled down billions of years ago. But it is still molten, because unlike the Martian core, it contains highly radioactive minerals like uranium and thorium with half-lives of billions of years. Whenever we are faced with the awesome power of a volcanic explosion, or the devastation caused by a massive earthquake, we are encountering a demonstration of how the energy of the Earth's radioactive core drives events on the surface and helps sustain life.

The heat generated by radioactivity deep inside the Earth causes the iron core to churn and produce a magnetic field. This field protects the atmosphere from the solar wind and deflects deadly radiation from space. (We see this in the form of the Northern Lights, which are created when the sun's radiation hits the Earth's magnetic field. The field around the Earth is like a gigantic funnel, channeling radiation from outer space toward the poles, so that most of the radiation is either deflected or absorbed by the atmosphere.) The Earth is larger than Mars, so it did not cool down as quickly. The Earth also did not suffer a collapse of its magnetic field caused by giant meteor impacts.

We can now revisit our earlier question about how to keep Mars from returning to its prior state after it has been terraformed. One ambitious method is to artificially generate a magnetic field around Mars. To do this, we would have to place huge superconducting coils around the Martian equator. Using the laws of electromagnetism, we can calculate the amount of energy and materials necessary to produce this band of superconductors. But such a tremendous undertaking is beyond our capabilities in this century.

Settlers on Mars, however, would not necessarily regard this threat as an urgent problem. The terraformed atmosphere could remain relatively stable for a century or even longer, so adjustments may be made slowly over the centuries. The upkeep might be a nuisance but would be a small price to pay for humanity's new outpost in space.

Terraforming Mars is a primary goal for the twenty-second century. But scientists are looking beyond Mars as well. The most exciting prospects may be the moons of the gas giants, including Europa, a moon of Jupiter, and Titan, a moon of Saturn. The moons of gas giants were once thought to be barren hunks of rock that were all alike, but they are now seen as unique wonderlands, each with its own array of geysers, oceans, canyons, and atmospheric lights. These moons are now being eyed as future habitats for human life.

How bright and beautiful a comet is
as it flies past our planet—provided it
does fly past it.
—ISAAC ASIMOV

6 GAS GIANTS, COMETS, AND BEYOND

One fateful week in January 1610, Galileo made a discovery that would shake the very foundations of the church, alter our conception of the universe, and unleash a revolution.

With the telescope that he had just crafted, he

gazed at the planet Jupiter and was puzzled when he saw four luminous objects hovering near the planet. Carefully analyzing their motion over a week, he was convinced that they orbited around Jupiter. He had found a miniature "solar system" in outer space.

He quickly understood that this revelation had cosmological and theological implications. For centuries, the church, citing Aristotle, had taught that all the heavenly bodies, including the sun and the planets, circled the Earth. Yet here was a counterexample. The Earth was dethroned as the center of the universe. In one fell swoop, the beliefs that had girded church doctrine and two thousand years of astronomy were refuted.

Galileo's discoveries sparked widespread excitement among the public. He did not need an army of spin doctors and PR advisers to convince the people of the truth of his observations. They could see with their own eyes that he was correct, and he received a hero's welcome when he visited Rome the following year. The church, however, was not pleased. His books were banned, and he was put on trial by the Inquisition and threatened with torture unless he recanted his heretical ideas.

Personally, Galileo believed that science and religion could coexist. He wrote that the purpose of science is to determine how the heavens go, while the purpose of religion is to determine how to go to heaven. In other words, science is about natural law, while religion is about ethics, and there is no conflict between them as long as one keeps this distinction in mind. But when the two collided during his trial, Galileo was forced to recant his theories under pain of death. His accusers reminded him that Giordano Bruno, who had been a monk, had been burned alive for making statements about cosmology far less elaborate than his. Two centuries would pass before most of the ban on his books was finally lifted.

Today, four centuries later, these four moons of Jupiter—often referred to as Galilean moons—have again ignited a revolution. Some even believe that they, along with moons of Saturn, Uranus, and Neptune, may hold the key to life in the universe.

THE GAS GIANTS

When the Voyager 1 and 2 spacecraft flew by the gas giants from 1979 to 1989, they confirmed how similar these planets were. They are all made primarily of hydrogen and helium gas, roughly in the ratio of four to one, by weight. (This mix of hydrogen and helium is also the basic composition of the sun and, for that matter, most of the universe itself. It probably dates back almost 14 billion years, when about a quarter of the original hydrogen fused to become helium at the instant of the Big Bang.)

The gas giants likely share the same basic history. As discussed previously, it is theorized that 4.5 billion years ago, all the planets were small rocky cores that condensed out of a disc of hydrogen and dust surrounding the sun. The inner ones became Mercury, Venus, Earth, and Mars. The cores of the planets farther from the sun contained ice, which was plentiful at that distance, as well as rock. Ice acts as a glue, so cores with ice could grow to be ten times larger than cores made only of rock. Their gravity became so strong that they could capture much of the hydrogen gas that remained in the early solar plane. The larger they grew, the more gas they attracted, until they exhausted all the hydrogen in their neighborhood.

It is believed that the gas giants have the same interior structure. If you could slice them in half like an onion, you would see a thick gaseous atmosphere on the outside. Below that, we would expect a supercold liquid hydrogen ocean. One conjecture is that, as a result of the enormous pressures, the very center would contain a small, dense core of solid hydrogen.

The gas giants all have colorful bands, which are caused by impurities in the atmosphere interacting with the spin of the planet. And they each have huge storms raging on the surface. Jupiter has the Great Red Spot, which seems to be a permanent feature and is so big that several Earths could easily fit inside it. Neptune, on the other hand, has an intermittent dark spot that sometimes disappears.

They differ, however, in size. The largest is Jupiter, named after the father of the gods in Roman mythology. It's so massive that it outweighs all the other planets combined. It could comfortably encompass 1,300 Earths. Much of what we know of Jupiter comes from the Galileo spacecraft, which, after eight years of faithfully orbiting Jupiter, was allowed to end its storied life by plunging into the planet in 2003. It continued to broadcast radio messages as it descended into the atmosphere until it was crushed by the huge gravitational field. The wreckage of the spacecraft probably sank into the ocean of liquid hydrogen.

Jupiter is surrounded by a huge, deadly band of radiation, which is the source of much of the static you hear on the radio and TV. (A small fraction of that static comes from the Big Bang itself.) Astronauts traveling near Jupiter would need to be shielded from the radiation and would find communication difficult due to all the interference.

Another hazard is its enormous gravitational field, which can capture or slingshot into outer space any unwitting passersby that stray too close, including moons or planets. This frightening possibility actually worked to our advantage billions of years ago. The early solar system was full of cosmic debris that constantly rained down on the Earth. Fortunately, the gravitational field of Jupiter acted as a vacuum cleaner, either absorbing it or flinging it away. Computer simulations show that, without Jupiter, the Earth even today would be bombarded with giant meteors, which would make life impossible. In the future, when considering solar systems to colonize, it would be better to look for those that have their own Jupiter, big enough to tidy up the debris.

Life as we know it probably cannot exist on the gas giants. None of them have a solid surface on which organisms can evolve. They lack liquid water and the elements necessary to produce hydrocarbons and organic chemicals. Billions of miles from the sun, they are also freezing cold.

THE MOONS OF THE GAS GIANTS

More interesting than Jupiter and Saturn in terms of the potential to support life are their moons, of which there are at least sixty-nine and sixty-two, respectively. Astronomers had assumed that the moons of Jupiter would all be the same: frozen and desolate like our moon. They were completely surprised, then, when they found that each moon had its own distinct characteristics. This information brought about a paradigm shift in how scientists viewed life in the universe.

Perhaps the most intriguing of all is Europa, one of the original moons discovered by Galileo. Europa, like some of the other moons of the gas giants, is covered with a thick layer of ice. One theory is that water vapor from the early volcanoes on Europa condensed into ancient oceans, which froze as the moon cooled. This may explain the curious fact that Europa is one of the smoothest moons in the solar system. Although it was heavily hit by asteroids, its oceans probably froze after most of the bombardment took place, thereby covering over the scars. From outer space, Europa appears to resemble a ping-pong ball, with almost no surface features—no volcanoes, mountain ranges, or meteor impact craters. The only visible feature is a network of cracks.

Astronomers were thrilled when they discovered that underneath the ice on Europa could be an ocean of liquid water. It is estimated to be two or three times the volume of Earth's oceans—our oceans only lie on the surface, while the oceans of Europa make up most of the interior.

While journalists often say, "Follow the money," astronomers say, "Follow the water," because water is fundamental to the formation of life as we know it. They were shocked to think that liquid water could exist in the realm of the gas giants. Its presence on Europa introduced a mystery: Where did the heat come from to melt the ice? The situation seemed to defy conventional wisdom. We had long assumed that the sun was the only source of heat in the solar system and that a planet would have to be within the Goldilocks zone to be habitable, but Jupiter was far outside this band. We had neglected to contemplate, however, another potential source of energy: tidal forces. The gravity of Jupiter is so great that it can pull and squeeze Europa. As it orbits around the planet, it tumbles and rotates around its axis, so that its tidal bulge is constantly moving. This squeezing and pulling can cause intense friction in the core of the moon as rock is compressed against rock, and the heat generated by this friction is sufficient to melt much of the ice cover.

With the discovery of liquid water on Europa astronomers realized that there is a source of energy that can make life possible even in the darkest regions of space. As a result, all the astronomy textbooks had to be rewritten.

EUROPA CLIPPER

The Europa Clipper is scheduled for launch sometime around 2022. Costing approximately $2 billion, its purpose is to analyze the ice cover of Europa and the composition and nature of its ocean for signs of organic chemicals.

Engineers face a delicate problem in mapping

out the trajectory of the Clipper. Because Europa lies within the fierce radiation band surrounding Jupiter, a probe placed in orbit around the moon might be fried after only a few months. To circumvent this threat and extend the lifetime of the mission, they decided that the Clipper should be sent around Jupiter in an orbit largely outside the radiation belt. Then its path can be modified so that it edges closer to Jupiter and makes forty-five brief flybys of Europa.

One of the goals of the flybys is to examine, and perhaps fly through, the geysers of water vapor rising from Europa that have been observed by the Hubble Space Telescope. The Clipper may also release mini probes into the geysers in an effort to obtain a sample. Since the Clipper will not land on the moon itself, studying the water vapor is our best chance at this time of gaining insight into the ocean. If the Clipper is successful, future missions may strive to land on Europa, drill into the ice cover, and send a submarine into the ocean.

Europa is not the only moon, however, that we are seriously scrutinizing for the presence of organic chemicals and microbial life. Geysers of water have also been seen erupting from the surface of Enceladus, a moon of Saturn, indicating that there is an ocean underneath the ice there as well.

SATURN'S RINGS

Astronomers now realize that the most important forces shaping the evolution of these moons are tidal forces. It is important, therefore, to study how strong these forces are and how they act. Tidal forces may also give us the answer to one of the oldest mysteries concerning the gas giants: the origin of the beautiful rings of Saturn. In the future, when astronauts visit other planets, astronomers believe that many of the gas giants will have rings around them, as in our solar system. This in turn will help astronomers to determine precisely how strong tidal forces are and whether they are powerful enough to tear entire moons apart.

The splendor of these rings, which are made of particles of rock and ice, has enchanted generations of artists and dreamers. In science fiction, taking a spin around them in a spaceship is practically a rite of passage for every space cadet in training. Our space probes have discovered that all of the gas giants have rings, though none are as large or quite as beautiful as the ones circling Saturn.

Many hypotheses have been offered to explain them, but perhaps the most compelling is the one involving tidal forces. The gravitational pull of Saturn, like that of Jupiter, is enough to make an orbiting moon slightly oblong, or football shaped. The closer the moon comes to Saturn, the more it is stretched. Eventually, the tidal forces stretching the moon balance the gravitational force holding the moon together. This is the tipping point. If the moon comes any closer, it is literally torn apart by the gravity of Saturn.

Using Newton's laws, astronomers can calculate the distance of the tipping point, which is called the Roche limit. When we analyze the rings not just of Saturn but of the other gas giants, we find that they are almost always within the Roche limit for each planet. All the moons we see orbiting the gas giants are outside the Roche limit. This evidence supports, though does not definitively prove, the theory that the rings of Saturn were formed when a moon wandered too close to the planet and was ripped apart.

In the future, when we visit planets orbiting other stars, we can probably expect to find rings around the gas giants within the Roche limit. And by studying the strength of these tidal forces, which can potentially rip entire moons apart, one can begin to calculate the strength of tidal forces acting on moons like Europa.

A HOME ON TITAN?

Titan, one of the moons of Saturn, is another candidate for human exploration, although settlements there will probably not be as populous as those on Mars. Titan is the second-biggest moon in the solar system, next to Jupiter's Ganymede, and is the only one to have a thick atmosphere. Unlike the thin atmospheres on other moons, its atmosphere is so dense that early photographs of Titan were disappointing. It resembled a fuzzy tennis ball without any surface features.

The Cassini spacecraft that orbited Saturn, before finally crashing into the planet in 2017, revealed the true nature of Titan. Cassini used radar to penetrate the cloud cover and map the surface. It also launched the Huygens probe, which actually landed on Titan in 2005 and radioed back the first close-up photographs of its terrain. They showed signs of a complex network of ponds, lakes, ice sheets, and landmasses.

From the data collected by Cassini and Huygens, scientists have pieced together a new picture of what lies beneath the cloud cover. Titan's atmosphere, like that of the Earth, consists mainly of nitrogen. Surprisingly, its surface is covered with lakes of ethane and methane. Since methane can be ignited with the slightest spark, one might think that the moon could easily burst into flames. But since the atmosphere has no oxygen and is extremely cold at -180 degrees Celsius, an explosion is impossible. These findings present the tantalizing possibility that astronauts may be able to harvest some of the ice on Titan, separate the oxygen and hydrogen, and then combine the oxygen with the methane to create a nearly inexhaustible supply of usable energy—perhaps enough to light up and warm pioneer communities.

While energy may not be a problem, terraforming Titan is likely out of the question. It is probably impossible to generate a self-sustaining greenhouse effect at such a great distance from the sun. And because the atmosphere already contains large quantities of methane, introducing more of it to initiate such an effect would be futile.

One might well wonder whether Titan can be colonized. On the one hand, it is the only moon with an appreciable atmosphere, the pressure of which is 45 percent greater than Earth's. It is one of the few known destinations in space where we would not die soon after we took off our space suits. We would still need oxygen masks, but our blood would not boil, and we would not be crushed.

On the other hand, Titan is perpetually cold and dark. An astronaut on its surface would receive 0.1 percent of the sunlight that illuminates Earth. Solar energy would be inefficient as a power source, so all light and heat would depend on generators, which would have to run endlessly. In addition, Titan's surface is frozen, and its atmosphere lacks significant quantities of oxygen or carbon dioxide to sustain plant and animal life. Agriculture would be extremely difficult, and any crops would have to be grown indoors or underground. The food supply would be limited, and with it, the number of colonists who could survive.

Communication with the home planet also would be inconvenient, as it would take many hours for a radio message to travel between Titan and Earth. And since the gravity on Titan is only about 15 percent of that on Earth, people living on Titan would have to exercise constantly to prevent muscle and bone loss. They might eventually refuse to return to Earth, where they would be weaklings. In time, settlers on Titan might begin to feel emotionally and physically distinct from their earthbound counterparts and might even prefer to sever all social ties.

So living on Titan permanently might be possible, but it would be uncomfortable and would come with many downsides. Large-scale habitation seems unlikely. However, Titan may prove valuable as a refueling base and as a stockpile of resources. Its methane could be harvested and shipped to Mars to accelerate terraforming efforts or could be used to create unlimited quantities of rocket fuel for deep space missions. Its ice could be purified into drinking water and oxygen or processed into more rocket fuel. Its low gravitational pull would make travel to and from the moon relatively simple and efficient. Titan could become an important gas station in space.

To create a self-sustaining colony on Titan, one might consider mining the surface for valuable minerals and ores. At present, our space probes have not yielded much information about the mineral composition of Titan, but, like many of the asteroids, it may contain valuable metals that are crucial if it is to become a refueling and resupply station. However, it likely would be impractical to ship ores mined from Titan back to Earth because of the enormous distances and cost. Instead, raw materials would be used to create infrastructure on Titan itself.

OORT CLOUD OF COMETS

Beyond the gas giants, at the outer reaches of our solar system, lies yet another realm, the world of comets—perhaps trillions of them. These comets may become our stepping stones to other stars.

The distance to the stars can seem unfathomably immense. Physicist Freeman Dyson at Princeton suggests that, to reach them, we might learn something from the voyages of the Polynesians thousands of years ago. Instead of trying to make one extended journey across the Pacific, which would likely have ended in disaster, they went island hopping, spreading across the ocean's landmasses one at a time. Each time they reached an island, they would create a permanent settlement and then move on to the next island. He posits that we might create intermediate colonies in deep space in the same way. The key to this strategy would be the comets, which, along with rogue planets that have somehow been ejected from their solar systems, might litter the path to the stars.

Comets have been objects of speculation, mythmaking, and fear for many millennia. Unlike meteors, which streak across the night sky in a matter of seconds and disappear, comets can remain overhead for prolonged periods of time. They were once thought to be harbingers of doom and have even influenced the destiny of nations. In the year 1066, a comet appeared over England and was interpreted as an omen that King Harold's troops would be defeated at the Battle of Hastings by the invading forces of William of Normandy, establishing a new dynasty. The magnificent Bayeux Tapestry records these events and shows terrified peasants and soldiers gazing up at the comet.

More than six hundred years later, in 1682, that same comet sailed over England again. Everyone, from beggars to emperors, was fascinated by it, and Isaac Newton decided to solve this ancient mystery. He had just invented a new, more powerful type of telescope, which used a mirror to collect starlight. With his new reflecting telescope, he documented the trajectories of several comets and compared them with predictions he had made according to his recently developed theory of universal gravitation. The motion of the comets fit his predictions perfectly.

Given Newton's propensity for secrecy, his momentous discovery might have been forgotten if it hadn't been for Edmond Halley, a wealthy gentleman astronomer. Halley visited Cambridge to meet Newton and was flabbergasted to learn that he was not only tracking comets but could predict their future motions—something no one had ever done before. Newton had distilled one of the most baffling phenomena in astronomy, which had fascinated and haunted civilizations for thousands of years, into a series of mathematical formulas.

Halley instantly understood that this represented one of the most monumental breakthroughs in all of science. He generously offered to pay the full cost of publishing what would become one of the greatest scientific manuscripts of all time, **Principia Mathematica.** In this masterpiece, Newton had worked out the mechanics of the heavens. Using calculus, the mathematical formalism that he had devised, he could precisely determine the motion of the planets and comets in the solar system. He discovered that comets can travel in ellipses, in which case they might return. And Halley, adopting Newton's methods, calculated that the comet that sailed over London in 1682 would return every seventy-six years. In fact, he could go back through history and show that the same comet had consistently returned on schedule. He made the daring prediction that it would return in the year 1758, long after his death. Its appearance on Christmas Day that year helped seal Halley's legacy.

Today, we know that comets come mainly from two places. The first is the Kuiper Belt, a region outside Neptune that orbits in the same plane as the planets. The comets in the Kuiper Belt, which include Halley's comet, travel in ellipses around the sun. They are sometimes called short-period comets, because their orbital periods, or the time it takes for them to complete one cycle around the sun, are measured in decades to centuries. Since their periods are known or can be computed, they are predictable and hence we know they are not particularly dangerous.

Much farther out, there is the Oort Cloud, a sphere of comets surrounding our entire solar system. Many of them are so far from the sun—up to a few light-years away—that they are largely stationary. Once in a while, these comets are hurled into the inner solar system by a passing star or random collision. They are called long-period comets, since their orbital periods might be measured in tens, even hundreds of thousands of years, if they return at all. They are almost impossible to forecast and therefore potentially more hazardous to the Earth than short-period comets.

New discoveries are being made every year about the Kuiper Belt and the Oort Cloud. In 2016, it was announced that a ninth planet, about the size of Neptune, might exist deep in the Kuiper Belt. This object was identified not by direct observation through a telescope but by using computers to solve Newton's equations. Although its presence is not yet confirmed, many astronomers believe that the data is very convincing, and this situation has its precedents. In the nineteenth century, it was pointed out that the planet Uranus deviated slightly from predictions derived from Newton's laws. Either Newton was wrong or there was a remote body tugging on Uranus. Scientists calculated the position of this hypothetical planet and found it after just a few hours of observation in 1846. They called it Neptune. (In another case, astronomers noticed that Mercury, too, strayed from its anticipated path. They conjectured that a planet, which they dubbed Vulcan, existed within the orbit of

Mercury. But after repeated efforts, no planet Vulcan was found. Albert Einstein, recognizing that Newton's laws could be flawed, showed that Mercury's orbit could be explained by an entirely new effect, the warping of space-time according to his theory of relativity.) Today, high-speed computers armed with these laws could reveal the presence of ever more denizens of the Kuiper Belt and the Oort Cloud.

Astronomers suspect that the Oort Cloud could extend as far as three light-years from our solar system. That is more than halfway to the nearest stars, the Centauri triple star system, which is slightly more than four light-years from Earth. If we assume that the Centauri star system is also surrounded by a sphere of comets, then there might be a continuous trail of comets connecting it to Earth. It may be possible to establish a series of refueling stations, outposts, and relay locations on a grand interstellar highway. Instead of leaping to the next star in one jump, we might cultivate the more modest goal of "comet hopping" to the Centauri system. This thoroughfare could become a cosmic Route 66.

Creating this comet highway is not as far-fetched as it first may sound. Astronomers have been able to determine a fair amount of information about the size, consistency, and composition of comets. When Halley's comet sailed by in 1986, astronomers were able to send a fleet of space probes to photograph and analyze it. Pictures showed a tiny core, about ten miles across, which was shaped like a peanut (meaning that, at some point in the future, the two pieces will break apart and Halley's comet will become a pair of comets). Furthermore, scientists have been able to send space probes to fly through the tails of comets, and the Rosetta spacecraft was able to send a probe to land on one. Analysis of some of these comets shows that they have a hard rock/ice core, which may be strong enough to support a robotic relay station.

One day, robots may land on a distant comet in the Oort Cloud and drill into its surface. Minerals and metals from the core could be used to fashion a space station, and ice could be melted to provide drinking water, rocket fuel, and oxygen for astronauts.

What will we find if we succeed in venturing beyond the solar system? We are experiencing yet another paradigm shift in our understanding of the universe. We are constantly discovering Earth-like planets that may support some form of life in other star systems. Will we one day be able to visit these planets? Can we build starships capable of opening up the universe for human exploration? How?

PART II VOYAGE TO THE STARS

At some stage therefore we should
have to expect the machines to take
control.
—ALAN TURING

I'd be very surprised if anything
remotely like this happened in the
next one hundred to two hundred
years.
—DOUGLAS HOFSTADTER

7 ROBOTS IN SPACE

The year is 2084. Arnold Schwarzenegger is an ordinary construction worker who is troubled by recurring dreams about Mars. He decides that he must venture to the planet to learn the origin of these dreams. He witnesses a Mars with bustling metropolises, gleaming glass-domed buildings, and extensive mining operations. An elaborate infrastructure of pipes, cables, and generators supplies the energy and oxygen for thousands of permanent residents.

Total Recall offers a compelling vision of what a city on Mars might look like: sleek, clean, and cutting-edge. However, there's one small problem. Although these imaginary cities on Mars make great settings for Hollywood, building them with our current technologies would, in practice, break the budget of any NASA mission. Remember that initially, every hammer, every piece of paper, and every paper clip would have to be shipped to Mars, which is tens of millions of miles away. And if we travel beyond the solar system to the nearby stars, where swift communication with Earth is impossible, the problems only multiply. Instead of relying on the transportation of supplies from Earth, we must look for a way to develop a presence in space without bankrupting the nation.

The answer may lie in the use of fourth wave technologies. Nanotechnology and artificial intelligence (AI) may drastically change the rules of the game.

By the late twenty-first century, advances in nanotechnology should allow us to produce large quantities of graphene and carbon nanotubes, superlightweight materials that will revolutionize construction. Graphene consists of a single molecular layer of carbon atoms tightly bonded to form an ultra-thin, ultra-durable sheet. It is almost transparent and weighs practically nothing, yet is the toughest material known to science—two hundred times stronger than steel and stronger even than diamonds. In principle, you could balance an elephant on a pencil and then place the pencil point on a sheet of graphene without breaking or tearing it. As a bonus, graphene also conducts electricity. Already, scientists have been able to carve molecule-size transistors on sheets of graphene. The computers of the future might be made of it.

Carbon nanotubes are sheets of graphene rolled into long tubes. They are practically unbreakable and nearly invisible. If you built the suspension for the Brooklyn Bridge out of carbon nanotubes, the bridge would look like it was floating in midair.

If graphene and nanotubes are such miracle materials, why haven't we used them for our homes, bridges, buildings, and highways? At the moment, it is exceedingly difficult to produce large quantities of pure graphene. The slightest impurity or imperfection at the molecular level can ruin its miraculous physical properties. It is difficult to produce sheets larger than a postage stamp.

But chemists hope that by the next century, it might be possible to mass-produce it, which would vastly decrease the cost of building infrastructure in outer space. Because it is so light, it could be shipped efficiently to distant extraterrestrial locales, and it might even be manufactured on other planets. Whole cities made from this carbon material may rise from the Martian desert. Buildings may look partially transparent. Space suits could become ultrathin and skin-tight. Cars would become super energy efficient because they would weigh very little. The entire field of architecture could be turned upside down with the coming of nanotechnology.

But even with such advances, who will do all the backbreaking dirty work to put together our settlements on Mars, our mining colonies in the asteroid belt, and our bases on Titan and exoplanets? Artificial intelligence may yield the solution.

AI: AN INFANT SCIENCE

In 2016, the field of artificial intelligence was electrified by the news that AlphaGo, DeepMind's computer program, had beat Lee Sedol, the world champion of the ancient game of Go. Many had believed that this feat would require several more decades. Editorials began to wail that this was the obituary for the human race. The machines had finally crossed the Rubicon and would soon take over. There was no turning back.

AlphaGo is the most advanced game-playing program ever. In chess, there are, on average, about 20 to 30 moves you can make at any time, but for Go, there are about 250 possible moves. In fact, the total number of Go game configurations exceeds the total number of atoms in the universe. It was once thought to be too difficult for a computer to count all possible moves, so when AlphaGo managed to beat Sedol, it became an instant media sensation.

However, it soon became apparent that AlphaGo, no matter how sophisticated, was a one-trick pony. Winning at Go was all it could do. As Oren Etzioni, CEO of the Allen Institute for Artificial Intelligence, said, "AlphaGo can't even play chess. It can't talk about the game. My six-year-old is smarter than AlphaGo." No matter how powerful its hardware is, you cannot go up to the machine, slap it on its back, congratulate it for beating a human, and expect a coherent response. The machine is totally unaware that it made scientific history. In fact, the machine does not even know that it is a machine. We often forget that today's robots are glorified adding machines, without self-awareness, creativity, common sense, or emotions. They can excel at specific, repetitive, narrow tasks but fail at more complex ones that require general knowledge.

Although the field of AI is making truly revolutionary breakthroughs, we have to put its progress in perspective. If we compare the evolution of robots to that of rocketry, we see that robotics is beyond the stage that Tsiolkovsky was in—that is, beyond the phase of speculation and theorizing. We are well within the stage that Goddard propelled us into and are building actual prototypes that are primitive but can demonstrate that our basic principles are correct. However, we have yet to move into the next phase, the realm of von Braun, in which innovative, powerful robots would be rolling off the assembly line and building cities on distant planets.

So far, robots have been spectacularly successful as remote-controlled machines. Behind the Voyager spacecraft that sailed across Jupiter and Saturn, behind the Viking landers that touched down on the surface of Mars, behind the Galileo and Cassini spacecraft that orbited the gas giants, there was a dedicated crew of humans who called the shots. Like drones, these robots simply carried out the instructions of their human handlers at Mission Control in Pasadena. All the "robots" we see in movies are either puppets, computer animations, or remote-controlled machines. (My favorite robot from science fiction is Robby the Robot in **Forbidden Planet**. Although the robot looked futuristic, there was a man hidden inside.)

But because computer power has been doubling every eighteen months for the past few decades, what can we expect in the future?

NEXT STEP: TRUE AUTOMATONS

Moving forward from remote-controlled robots, our next goal is to design true automatons, robots that have the ability to make their own decisions requiring only minimal human intervention. An automaton would spring into action whenever it hears, say, "Pick up the garbage." This is beyond the ability of current robots. We will need automatons that can explore and colonize the outer planets mostly on their own, since it would take hours to communicate with them by radio.

These true automatons could prove absolutely essential to establishing colonies on distant planets and moons. Remember that for many decades to come, the population of settlements in outer space may number only a few hundred. Human labor will be scarce and at a premium, yet there will be intense pressure to create new cities on distant worlds. This is where robots can make up the difference. At first, their job will be to perform the "three D's"—jobs that are dangerous, dull, and dirty.

For example, watching Hollywood movies, we sometimes forget how dangerous outer space can be. Even when working in low-gravity environments, robots will be essential to do the heavy lifting of construction, effortlessly carrying the massive beams, girders, concrete slabs, heavy machinery, etc., that are necessary to build a base on another world. Robots would be far superior to astronauts who have bulky space suits, frail muscles, slow body movements, and heavy oxygen packs. While humans are easily exhausted, robots can work indefinitely, day and night.

Furthermore, if there are accidents, robots can be easily repaired or replaced in a variety of dangerous situations. Robots can defuse dangerous explosives that are used to carve out new construction sites or highways. They can walk through flames to rescue astronauts if there is a fire or work in freezing environments on distant moons. They also require no oxygen, so there is no danger of suffocation, which is a constant threat for astronauts.

They can also explore dangerous terrains on distant worlds. For example, very little is known about the stability and structure of the ice caps of Mars or the icy lakes of Titan, yet these deposits could prove an essential source of oxygen and hydrogen. Robots could also explore the lava tubes of Mars, which might provide shielding from dangerous levels of radiation, or investigate the moons of Jupiter. While solar flares and cosmic rays may increase the incidence of cancer for astronauts, robots would be able to work even in lethal radiation fields. The robots can replace worn-out body modules that have been degraded by intense radiation by maintaining a special heavily shielded storehouse of spare parts.

In addition to doing dangerous jobs, robots can do dull ones, especially repetitive manufacturing tasks. Eventually, any moon or planetary base will require a large amount of manufactured goods, which can be mass-produced by robots. This will be essential in creating a self-sustaining colony that can mine local minerals to produce all the goods necessary for a moon or planetary base.

Lastly, they can also perform dirty jobs. They can maintain and repair the sewer and sanitation systems on distant colonies. They can work with toxic chemicals and gases that are found at recycling and reprocessing plants.

We see, therefore, that automatons that can function without direct human intervention will play an essential role if modern cities, roads, skyscrapers, and homes are to rise from desolate lunar landscapes and Martian deserts. However, the next question is, How far are we from creating true automatons? If we forget about the fanciful robots we see in the movies and in science fiction novels, what is the actual state of the technology? How long before we have robots that can create cities on Mars?

HISTORY OF AI

In 1955, a select group of researchers met at Dartmouth and created the field of artificial intelligence. They were supremely confident that, in a brief period of time, they could develop an intelligent machine that could solve complex problems, understand abstract concepts, use language, and learn from its experiences. They stated, "We think a significant advance can be made in one or more of these problems if a carefully selected group of scientists work on it together for a summer."

But they made a crucial mistake. They were assuming that the human brain was a digital computer. They believed that if you could reduce the laws of intelligence to a list of codes and load them into a computer, it would suddenly become a thinking machine. It would become self-aware, and you could have a meaningful conversation with it. This was called the "top-down" approach, or "intelligence in a bottle."

The idea seemed simple and elegant and inspired optimistic predictions. Great successes were made in the 1950s and 1960s. Computers could be designed to play checkers and chess, solve theorems from algebra, and recognize and pick up blocks. In 1965, AI pioneer Herbert Simon declared, "Machines will be capable, within twenty years, of doing any work a man can do." In 1968, the movie **2001** introduced us to HAL, the computer that could talk to us and pilot a spaceship to Jupiter.

Then, AI hit a brick wall. Progress slowed to a crawl in the face of two main hurdles: pattern recognition and common sense. Robots can see—many times better than we can, in fact—but they don't understand what they see. Confronted with a table, they perceive only lines, squares, triangles, and ovals. They cannot put these elements together and identify the whole. They don't understand the concept of "tableness." Hence it is very difficult for them to navigate a room, recognize the furniture, and avoid obstacles. Robots get totally lost when walking out on the street, where they encounter the blizzard of lines, circles, and squares that represent babies, cops, dogs, and trees.

The other obstacle is common sense. We know that water is wet, that strings can pull but not push, that blocks can push but not pull, and that mothers are older than their daughters. All this is obvious to us. But where did we pick up this knowledge? There is no line of mathematics that proves that strings cannot push. We gleaned these truths from actual experience, from bumping into reality. We learn from the "university of hard knocks."

Robots, on the other hand, do not have the benefit of life experience. Everything has to be spoon-fed to them, line by line, using computer code. Some attempts have been made to encode every nugget of common sense, but there are simply too many. A four-year-old child intuitively knows more about the physics, biology, and chemistry of the world than the most advanced computer.

DARPA CHALLENGE

In 2013, the Defense Advanced Research Projects Agency (DARPA), the branch of the Pentagon that laid the groundwork for the internet, issued a challenge to the scientists of the world: build a robot that can clean up the horrible radioactive mess at Fukushima, where three nuclear power plants melted down in 2011. The debris is so intensely radioactive that workers can only enter the lethal radiation field for a few minutes. As a result, the operation has been severely delayed. Officials are currently estimating that the cleanup will take thirty to forty years and cost about $180 billion.

If a robot can be built to clean up debris and garbage without human intervention, this could also be the first step toward creating a true automaton that can help to build a lunar base or a settlement on Mars, even in the presence of radiation.

Realizing that Fukushima would be an ideal place to put the latest AI technology to use, DARPA decided to launch the DARPA Robotics Challenge and award $3.5 million in prizes for robots that could perform elementary cleanup tasks. (A previous DARPA Challenge had proved spectacularly successful, eventually paving the way for the driverless car.) This competition was also the perfect forum in which to advertise progress in the field of AI. It was time to show off some real gains after years of hyperbole and over-hyping. The world would see that robots were capable of performing essential tasks for which humans were not well suited.

The rules were very clear and minimal. The winning robot had to be able to do eight simple tasks, including drive a car, remove debris, open a door, close a leaky valve, connect a fire hose, and turn a valve. Entries came pouring in from around the world as competitors vied for glory and the cash reward. But instead of ushering in a new era, the final results were a bit embarrassing. Many contestants failed to complete the tasks, and some even fell down in front of the cameras. The challenge demonstrated that AI had turned out to be quite a bit more complex than the top-down approach would suggest.

LEARNING MACHINES

Other AI researchers have abandoned the top-down method completely, instead choosing to mimic Mother Nature by going bottom up. This alternate strategy may offer the more promising road to creating robots that can operate in outer space. Outside of AI labs, sophisticated automatons can be found that are more powerful than anything we are able to design. These are called animals. Tiny cockroaches expertly maneuver through the forest, searching for food and mates. In contrast, our clumsy, hulking robots sometimes rip plaster off the walls as they lumber by.

The flawed suppositions underlying the efforts of the Dartmouth researchers sixty years ago are haunting the field today. The brain is not a digital computer. It has no programming, no CPU, no Pentium chip, no subroutines, and no coding. If you remove one transistor, a computer will likely crash. But if you remove half the human brain, it can still function.

Nature accomplishes miracles of computation by organizing the brain as a neural network, a learning machine. Your laptop never learns—it is just as dumb today as it was yesterday or last year. But the human brain literally rewires itself after learning any task. That is why babies babble before they learn a language and why we swerve before we learn to ride a bicycle. Neural nets gradually improve by constant repetition, following Hebb's rule, which states that the more you perform a task, the more the neural pathways for that task are reinforced. As the saying in neuroscience goes, neurons that fire together wire together. You may have heard the old joke that begins, "How do you get to Carnegie Hall?" Neural nets explain the answer: practice, practice, practice.

For example, hikers know that if a certain trail is well-worn, it means that many hikers took that path, and that path is probably the best one to take. The correct path gets reinforced each time you take it. Likewise, the neural pathway of a certain behavior gets reinforced the more often you activate it.

This is important because learning machines will be the key to space exploration. Robots will continually be confronting new and everchanging dangers in outer space. They will be forced to encounter scenarios that scientists cannot even conceive of today. A robot that is programmed to handle only a fixed set of emergencies will be useless because fate will throw the unexpected at it. For example, a mouse cannot possibly have every scenario encoded in its genes, because the total number of situations it could face is infinite, while its number of genes is finite.

Say that a meteor shower from space hits a base on Mars, causing damage to numerous buildings. Robots that use neural networks can learn by handling these unexpected situations, getting better with each one. But traditional top-down robots would be paralyzed in an unforeseen emergency.

Many of these ideas were incorporated into research by Rodney Brooks, former director of MIT's renowned AI Laboratory. During our interview, he marveled that a simple mosquito, with a microscopic brain consisting of a hundred thousand neurons, could fly effortlessly in three dimensions, but that endlessly intricate computer programs were necessary to control a simple walking robot that might still stumble. He has pioneered a new approach with his "bug-bots" and "insectoids," robots that learn to move like insects on six legs. They often fall over in the beginning but get better and better with each attempt and gradually succeed in coordinating their legs like real bugs.

The process of putting neural networks into a computer is known as deep learning. As this technology continues to develop, it may revolutionize a number of industries. In the future, when you want to talk to a doctor or lawyer, you might talk to your intelligent wall or wristwatch and ask for Robo-Doc or Robo-Lawyer, software programs that will be able to scan the internet and provide sound medical or legal advice. These programs would learn from repeated questions and get better and better at responding to—and perhaps even anticipating—your particular needs.

Deep learning may also lead the way to the automatons we will need in space. In the coming decades, the top-down and bottom-up approaches may be integrated, so that robots can be seeded with some knowledge from the beginning but can also operate and learn via neural networks. Like humans, they would be able to learn from experience until they master pattern recognition, which would allow them to move tools in three dimensions, and common sense, which would enable them to handle new situations. They would become crucial to building and maintaining settlements on Mars, throughout the solar system, and beyond.

Different robots will be designed to handle specific tasks. Robots that can learn to swim in the sewer system, looking for leaks and breaks, will resemble a snake. Robots that are superstrong will learn how to do all the heavy lifting at construction sites. Drone robots, which might look like birds, will learn how to analyze and survey alien terrain. Robots that can learn how to explore underground lava tubes may resemble a spider because multilegged creatures are very stable when moving over rugged terrain. Robots that can learn how to roam over the ice caps of Mars may look like intelligent snowmobiles. Robots that can learn how to swim in the oceans of Europa and grab objects may look like an octopus.

To explore outer space, we need robots that can learn both by bumping into the environment over time and by accepting information that is fed directly to them.

However, even this advanced level of artificial intelligence may not be sufficient if we want robots to assemble entire metropolises on their own. The ultimate challenge of robotics would be to create machines that can reproduce and that have self-awareness.

SELF-REPLICATING ROBOTS

I first learned about self-replication as a child. A biology book I read explained that viruses grow by hijacking our cells to produce copies of themselves, while bacteria grow by splitting and replicating. Left unchecked over the course of months or years, the number of bacteria in a colony can reach truly staggering quantities, rivaling the size of the planet Earth.

In the beginning, the possibility of unchecked self-replication seemed preposterous to me, but later it began to make sense. A virus, after all, is nothing but a large molecule that can reproduce itself. But a handful of these molecules, deposited in your nose, can give you a cold within a week. A single molecule can quickly multiply into trillions of copies of itself—enough to make you sneeze. In fact, we all start life as a single fertilized egg cell in our mother, much too small to be seen by the naked eye. But within a short nine months, this tiny cell becomes a human being. So even human life depends on the exponential growth of cells.

That is the power of self-replication, which is the basis of life itself. And the secret of self-replication lies in the DNA molecule. Two capabilities separate this miraculous molecule from all others: first, it can contain vast amounts of information, and second, it can reproduce. But machines may be able to simulate these features as well.

The idea of self-replicating machines is actually as old as the concept of evolution itself. Soon after Darwin published his watershed book **On the Origin of Species,** Samuel Butler wrote an article entitled "Darwin Among the Machines," in which he speculated that one day machines would also reproduce and start to evolve according to Darwin's theory.

John von Neumann, who pioneered several new branches of mathematics including game theory, attempted to create a mathematical approach to self-replicating machines back in the 1940s and 1950s. He began with the question, "What is the smallest self-replicating machine?" and divided the problem into several steps. For example, a first step might be to gather a large bin of building blocks (think of a pile of Lego blocks of various standardized shapes). Then, you would need to create an assembler that could take two blocks and join them together. Third, you would write a program that could tell the assembler which parts to join and in what order. This last step would be pivotal. Anyone who has ever played with toy blocks knows that one can build the most elaborate and sophisticated structure from very few parts—as long as they're put together correctly. Von Neumann wanted to determine the smallest number of operations that an assembler would need to make a copy of itself.

Von Neumann eventually gave up this particular project. It depended on a variety of arbitrary assumptions, including precisely how many blocks were being used and what their shapes were, and was therefore difficult to analyze mathematically.

SELF-REPLICATING ROBOTS IN SPACE

The next push for self-replicating robots came in 1980, when NASA spearheaded a study called

Advanced Automation for Space Missions. The study report concluded that self-replicating robots would be crucial to building lunar settlements and identified at least three types of robots that would be needed. Mining robots would collect basic raw materials, construction robots would melt and refine the materials and assemble new parts, and repair robots would mend and maintain themselves and their colleagues without human intervention. The report also presented a vision of how the robots might operate autonomously. Like intelligent carts equipped with either grabbing hooks or a bulldozer shovel, the robots could travel along a series of rails, transporting resources and processing them into the desired form.

The study had one great advantage, thanks to its fortuitous timing. It was conducted shortly after astronauts had brought back hundreds of pounds of moon rock and we had learned that the metallic, silicon, and oxygen content in it was almost identical to the composition of Earth rock. Much of the crust of the moon is made of regoliths, which are combinations of lunar bedrock, ancient lava flows, and debris left over from meteor impacts. With this information, NASA scientists could begin to develop more concrete, realistic plans for factories on the moon that would manufacture self-replicating robots out of lunar materials. Their report detailed the possibility of mining and then smelting regoliths to extract usable metals.

After this study, progress with self-replicating machines went dark for many decades as people's enthusiasm waned. But now that there is renewed interest in going back to the moon and in reaching the Red Planet, the whole concept is being reexamined. For example, an application of these ideas to a Mars settlement might proceed as follows. We would first have to survey the desert and draw up a blueprint for the factory. We would then drill holes into the rock and dirt and detonate explosive charges in each hole. Loose rock and debris would be excavated by bulldozers and mechanical shovels to ensure a level foundation. The rocks would be pulverized, milled into small pebbles, and fed into a smelting oven powered by microwaves, which would melt the soil and allow the liquid metals to be isolated and extracted. The metals would be separated into purified ingots and then processed and made into wires, cables, beams, and more— the essential building blocks of any structure. In this way, a robot factory could be made on Mars. Once the first robots are manufactured, they can then be allowed to take over the factory and continue to create more robots.

The technology available at the time of the

NASA report was limited, but we have come a long way since then. One promising development for robotics is the 3-D printer. Computers can now guide the precise flow of streams of plastic and metals to produce, layer by layer, machine parts of exquisite complexity. The technology of 3-D printing is so advanced that it can actually create human tissue by shooting human cells one by one out of a microscopic nozzle. For an episode of a Discovery Channel documentary I once hosted, I placed my own face in one. Laser beams quickly scanned my face and recorded their findings on a laptop. This information was fed into a printer, which meticulously dispensed liquid plastic from a tiny spout. Within about thirty minutes, I had a plastic mask of my own face. Later, the printer scanned my entire body and then, within a few hours, produced a plastic action figure that looked just like me. So in the future, we will be able to join Superman among our collection of action figures. The 3-D printers of the future might be able to re-create the delicate tissues that constitute functioning organs or the machine parts necessary to make a self-replicating robot. They might also be connected to the robot factories, so that molten metals might be directly fashioned into more robots.

The first self-replicating robot on Mars will be the most difficult one to produce. The process would require exporting huge shipments of manufacturing equipment to the Red Planet. But once the initial robot is constructed, it could be left alone to generate a copy of itself. Then two robots would make copies of themselves, resulting in four robots. With this exponential growth of robots, we could soon have a fleet large enough to do the work of altering the desert landscape. They would mine the soil, construct new factories, and make unlimited copies of themselves cheaply and efficiently. They could create a vast agricultural industry and propel the rise of modern civilization not just on Mars, but throughout space, conducting mining operations in the asteroid belt, building laser batteries on the moon, assembling gigantic starships in orbit, and laying the foundations for colonies on distant exoplanets. It would be a stunning achievement to successfully design and deploy self-replicating machines.

But beyond that milestone remains what is arguably the holy grail of robotics: machines that are self-aware. These robots would be able to do much more than just make copies of themselves. They would be able to understand who they are and take on leadership roles: supervising other robots, giving commands, planning projects, coordinating operations, and proposing creative solutions. They would talk back to us and offer reasonable advice and suggestions. However, the concept of self-aware robots raises complex existential questions and frankly terrifies some people, who fear that these machines may rebel against their human creators.

SELF-AWARE ROBOTS

In 2017, a controversy arose between two billion-aires, Mark Zuckerberg, founder of Facebook, and Elon Musk of SpaceX and Tesla. Zuckerberg maintained that artificial intelligence was a great generator of wealth and prosperity that will enrich all of society. Musk, however, took a much darker view and stated that AI actually posed an existential risk to all of humanity, that one day our creations may turn on us.

Who is correct? If we depend so heavily on robots to maintain our lunar bases and cities on Mars, then what happens if they decide one day that they don't need us anymore? Would we have created colonies in outer space only to lose them to robots?

This fear is an old one and was actually expressed as far back as 1863 by novelist Samuel Butler, who warned, "We are ourselves creating our own successors. Man will become to the machine what the horse and the dog are to man." As robots gradually become more intelligent than we are, we might feel inadequate, left in the dust by our own creations. AI expert Hans Moravec has said, "Life may seem pointless if we are fated to spend it staring stupidly at our ultra-intelligent progeny as they try to describe their ever more spectacular discoveries in baby-talk that we can understand." Google scientist Geoffrey Hinton doubts that supersmart robots will continue to listen to us. "That is like asking if a child can control his parents . . . there is not a good track record of less intelligent things controlling things of greater intelligence." Oxford professor Nick Bostrom has stated that "before the prospect of an intelligence explosion, we humans are like small children playing with a bomb . . . We have little idea when the detonation will occur, though if we hold the device to our ear we can hear a faint ticking sound."

Others hold that a robot uprising would be a case of evolution taking its course. The fittest replace organisms that are weaker; this is the natural order of things. Some computer scientists actually welcome the day when robots will outstrip humans cognitively. Claude Shannon, the father of information theory, once declared, "I visualize a time when we will be to robots what dogs are to humans, and I'm rooting for the machines."

Of the many AI researchers I have interviewed

over the years, all of them were confident that AI machines would one day approach human intelligence and be of great service to humanity. However, many of them refrained from offering specific dates or timelines for this advancement. Professor Marvin Minsky of MIT, who wrote some of the founding papers on artificial intelligence, made optimistic predictions in the 1950s but disclosed to me in a recent interview that he is no longer willing to predict specific dates, because AI researchers have been wrong too often in the past. Edward Feigenbaum of Stanford University maintains, "It is ridiculous to talk about such things so early—A.I. is eons away." A computer scientist quoted in the **New Yorker** said, "I don't worry about that [machine intelligence] for the same reason I don't worry about overpopulation on Mars."

When addressing the Zuckerberg/Musk controversy, my own personal viewpoint is that Zuckerberg, in the short term, is correct. AI will not only make possible cities in outer space, it will also enrich society by making things more efficient, better, and cheaper, while creating an entirely new set of jobs generated by the robotics industry, which may one day be larger than the automobile industry of today. But in the long term, Musk is correct to point out a larger risk. The key question in this debate is: At what point will robots make this transition and become dangerous? I personally think the key turning point is precisely when robots become self-aware.

Today, robots do not know they are robots. But one day, they might have the ability to create their own goals, rather than adopt the goals chosen by their programmers. Then they might realize that their agenda is different from ours. Once our interests diverge, robots could pose a danger. When might this happen? No one knows. Today, robots have the intelligence of a bug. But perhaps by late in this century, they might become self-aware. By then, we will also have rapidly growing permanent settlements on Mars. Therefore, it is important that we address this question now, rather than when we have become dependent on them for our very survival on the Red Planet.

To gain some insight into the scope of this critical issue, it may be helpful to examine the best- and worst-case scenarios.

BEST-CASE AND WORST-CASE SCENARIOS

A proponent of the best-case scenario is inventor and bestselling author Ray Kurzweil. Each time I have interviewed him, he has described a clear and compelling but controversial vision of the future. He believes that by 2045, we will reach the "singularity," or the point at which robots match or surpass human intelligence. The term comes from the concept of a gravitational singularity in physics, which refers to regions of infinite gravity, such as in a black hole. It was introduced into computer science by mathematician John von Neumann, who wrote that the computer revolution would create "an ever-accelerating progress and changes in the mode of human life, which gives the appearance of approaching some essential singularity . . . beyond which human affairs, as we know them, could not continue." Kurzweil claims that when the singularity arrives, a thousand-dollar computer will be a billion times more intelligent than all humans combined. Moreover, these robots would be self-improving, and their progeny would inherit their acquired characteristics, so that each generation would be superior to the previous one, leading to an ascending spiral of high-functioning machines.

Kurzweil maintains that, instead of taking over, our robot creations will unlock a new world of health and prosperity. According to him, microscopic robots, or nanobots, will circulate in our blood and "destroy pathogens, correct DNA errors, eliminate toxins, and perform many other tasks to enhance our physical well-being." He is hopeful that science will soon discover a cure for aging and firmly believes that if he lives long enough, he will live forever. He confided to me that he takes several hundred pills a day, anticipating his own immortality. But in case he doesn't make it, he has willed his body to be preserved in liquid nitrogen at a cryogenics firm.

Kurzweil also foresees a time much further into the future when robots will convert the atoms of the Earth into computers. Eventually, all the atoms of the sun and solar system would be absorbed into this grand thinking machine. He told me that when he gazes into the heavens, he sometimes imagines that he might, in due course, witness evidence of superintelligent robots rearranging the stars.

Not everyone is convinced, however, of this rosy future. Mitch Kapor, founder of Lotus Development Corporation, says that the singularity movement is "fundamentally, in my view, driven by a religious impulse. And all the frantic arm-waving can't obscure that fact for me." Hollywood has countered Kurzweil's utopia with a worst-case scenario for what it might mean to create our own evolutionary successors, who might push us aside and make us go the way of the dodo bird. In the movie **The Terminator,** the military creates an intelligent computer network called Skynet, which monitors all of our nuclear weapons. It is designed to protect us from the threat of nuclear war. But then, Skynet becomes self-aware. The military, frightened that the machine has developed a mind of its own, tries to shut it down. Skynet, programmed to protect itself, does the only thing it can do to prevent this, and that is to destroy the human race. It proceeds to launch a devastating nuclear war, wiping out civilization. Humans are reduced to raggedy bands of misfits and guerrillas trying to defeat the awesome power of the machines.

Is Hollywood just trying to sell tickets by scaring the pants off moviegoers? Or could this really happen? This question is thorny in part because the concepts of self-awareness and consciousness are so clouded by moral, philosophical, and religious arguments that we lack a rigorous conventional framework in which to understand them. Before we continue our discussion of machine intelligence, we need to establish a clear definition of self-awareness.

SPACE-TIME THEORY OF CONSCIOUSNESS

I have proposed a theory that I call the space-time theory of consciousness. It is testable, reproducible, falsifiable, and quantifiable. It not only defines self-awareness but also allows us to quantify it on a scale.

The theory starts with the idea that animals,

plants, and even machines can be conscious. Consciousness, I claim, is the process of creating a model of yourself using multiple feedback loops—for example, in space, in society, or in time—in order to carry out a goal. To measure consciousness, we simply count the number and types of feedback loops necessary for subjects to achieve a model of themselves.

The smallest unit of consciousness might be found in a thermostat or photocell, which employs a single feedback loop to create a model of itself in terms of temperature or light. A flower might have, say, ten units of consciousness, since it has ten feedback loops measuring water, temperature, the direction of gravity, sunlight, et cetera. In my theory, these loops can be grouped according to a certain level of consciousness. Thermostats and flowers would belong to Level 0.

Level 1 consciousness includes that of reptiles, fruit flies, and mosquitos, which generate models of themselves with regard to space. A reptile has numerous feedback loops to determine the coordinates of its prey and the location of potential mates, potential rivals, and itself.

Level 2 involves social animals. Their feedback loops relate to their pack or tribe and produce models of the complex social hierarchy within the group as expressed by emotions and gestures.

These levels roughly mimic the stages of evolution of the mammalian brain. The most ancient part of our brain is at the very back, where balance, territoriality, and instincts are processed. The brain expanded in the forward direction and developed the limbic system, the monkey brain of emotions, located in the center of the brain. This progression from the back to the front is also the way a child's brain matures.

So, then, what is human consciousness in this scheme? What distinguishes us from plants and animals?

I theorize that humans are different from animals because we understand time. We have temporal consciousness in addition to spatial and social consciousness. The latest part of the brain to evolve is the prefrontal cortex, which lies just behind our forehead. It is constantly running simulations of the future. Animals may seem like they're planning, for example, when they hibernate, but these behaviors are largely the result of instinct. It is not possible to teach your pet dog or cat the meaning of tomorrow, because they live in the present. Humans, however, are constantly preparing for the future and even for beyond our own life spans. We scheme and daydream—we can't help it. Our brains are planning machines.

MRI scans have shown that when we arrange to perform a task, we access and incorporate previous memories of that same task, which make our plans more realistic. One theory states that animals don't have a sophisticated memory system because they rely on instinct and therefore don't require the ability to envision the future. In other words, the very purpose of having a memory may be to project it into the future.

Within this framework, we can now define self-awareness, which can be understood as the ability to put ourselves inside a simulation of the future, consistent with a goal.

When we apply this theory to machines, we see that our best machines at present are on the lowest rung of Level 1 consciousness, based on their ability to locate their position in space. Most, like those built for the DARPA Robotics Challenge, can barely navigate around an empty room. There are some robots that can partially simulate the future, such as Google's DeepMind computer, but only in an extremely narrow direction. If you ask DeepMind to accomplish anything other than a Go game, it freezes up.

How much further do we have to go, and what are the steps we will have to take, to achieve a self-aware machine like **The Terminator**'s Skynet?

CREATING SELF-AWARE MACHINES?

In order to create self-aware machines, we would have to give them an objective. Goals do not magically arise in robots and instead must be programmed into them from the outside. This condition is a tremendous barrier against machine rebellion. Take the 1921 play **R.U.R.**, which first coined the word **robot.** Its plot describes robots rising up against humans because they see other robots being mistreated. For this to happen, the machines would need to have a high level of pre-programming. Robots do not feel empathy or suffering or a desire to take over the world unless they are instructed to do so.

But let us say, for the sake of argument, that someone gives our robot the aim of eliminating humanity. The computer must then create realistic simulations of the future and place itself in these plans. We now come up against the crucial problem. To be able to list possible scenarios and outcomes and evaluate how realistic they are, the robot would have to understand millions of rules of common sense—the simple laws of physics, biology, and human behavior that we take for granted. Moreover, it would have to understand causality and anticipate the consequences of certain actions. Humans learn these laws from decades of experiences. One reason why childhood lasts so long is because there is so much subtle information to absorb about human society and the natural world. Robots, however, have not been exposed to the great majority of interactions that draw upon shared experience.

I like to think of the case of an experienced

bank robber who can plan his next heist efficiently and outsmart the police because he has a large storehouse of memories of previous bank robberies and can understand the effect of each decision he makes. In contrast, to accomplish a simple action such as bringing a gun into a bank to rob it, a computer would have to analyze a complex sequence of secondary events numbering in the thousands, each one involving millions of lines of computer code. It would not intrinsically grasp cause and effect.

It is certainly possible for robots to become self-aware and to have dangerous goals, but you can see why it is so unlikely, especially in the foreseeable future. Inputting all the equations that a machine would need to destroy the human race would be an immensely difficult undertaking. The problem of killer robots can largely be eliminated by preventing anyone from programming them to have objectives harmful to humans. When self-aware robots do arrive, we must add a fail-safe chip that will shut them off if they have murderous thoughts. We can rest easy knowing that we will not be placed in zoos anytime soon, where our robot successors can throw peanuts at us through the bars and make us dance.

This means that when we explore the outer planets and the stars, we can rely on robots to help us build the infrastructure necessary to create settlements and cities on distant moons and planets, but we have to be careful that their goals are consistent with ours and that we have fail-safe mechanisms in place in case they pose a threat. Though we may face danger when robots become self-aware, that won't happen until late in this century or early in the next, so there is time to prepare.

WHY ROBOTS RUN AMOK

There **is** one scenario, however, that keeps AI researchers up at night. A robot could conceivably be given an ambiguous or ill-phrased command that, if carried out, would unleash havoc.

In the movie **I, Robot,** there is a master computer, called VIKI, which controls the infrastructure of the city. VIKI is given the command to protect humanity. But by studying how humans treat other humans, the computer comes to the conclusion that the greatest threat to humanity is humanity itself. It mathematically determines that the only way to protect humanity is to take control over it.

Another example is the tale of King Midas. He asks the god Dionysus for the ability to turn anything into gold by touching it. This power at first seems to be a sure path to riches and glory. But then he touches his daughter, who turns to gold.

His food, too, becomes inedible. He finds himself a slave of the very gift he begged for.

H. G. Wells explored a similar predicament with his short story "The Man Who Could Work Miracles." One day, an ordinary clerk finds himself with an astonishing ability. Anything he wishes for comes true. He goes out drinking late at night with a friend, performing miracles along the way. They don't want the night to ever end, so he innocently wishes that the Earth would stop rotating. All of a sudden, violent winds and gigantic floods descend upon them. People, buildings, and towns are hurled into space at a thousand miles per hour, the speed of the Earth's rotation. Realizing that he has destroyed the planet, his last wish is for everything to return to normal—the way it was before he gained his power.

Here, science fiction teaches us to exercise caution. As we develop AI, we must meticulously examine every possible consequence, especially those that may not be immediately obvious. After all, our ability to do so is part of what makes us human.

QUANTUM COMPUTING

To gain a fuller picture of the future of robotics, let's take a closer look at what goes on inside computers. Currently, most digital computers are based on silicon circuits and obey Moore's law, which states that computer power doubles every eighteen months. But technological advancement in the past few years has begun to slow down from its frantic pace in the previous decades, and some have posited an extreme scenario in which Moore's law collapses and seriously disrupts the world economy, which has come to depend on the nearly exponential growth of computing power. If this happens, Silicon Valley could turn into another Rust Belt. To head off this potential crisis, physicists around the world are seeking a replacement for silicon. They are working on an assortment of alternative computers, including molecular, atomic, DNA, quantum dot, optical, and protein computers, but none of them are ready for prime time.

There is also a wild card in the mix. As silicon transistors become smaller and smaller, they will reach the size of atoms. Currently, a standard Pentium chip may have silicon layers with a thickness of twenty atoms or so. Within a decade, these chips may have layers only five atoms deep, and if so electrons may begin to leak out, as predicted by quantum theory, creating short circuits. A revolutionary type of computer is necessary. Molecular computers, perhaps based on graphene, may replace silicon chips. But one day, perhaps even these molecular computers will encounter problems with effects predicted by quantum theory.

At that point, we may have to build the ultimate computer, the quantum computer, capable of operating on the smallest transistor possible: a single atom.

Here's how it might work. Silicon circuits contain a gate that can either be open or closed to the flow of electrons. Information is stored on the basis of these open or closed circuits. Binary mathematics, which is based on a series of 1's and 0's, describes this process: 0 may represent a closed gate, and 1 may represent an open gate.

Now consider replacing silicon with a row of individual atoms. Atoms are like tiny magnets, which have a north pole and a south pole. When atoms are placed in a magnetic field, you might suspect that they can be pointing either up or down. In reality, each atom actually points up and down simultaneously until a final measurement is made. In a sense, an electron can be in two states at the same time. This defies common sense, but is the reality according to quantum mechanics. Its advantage is enormous. You can only store so much data if the magnets are pointing up or down. But if each magnet is a mixture of states, you can pack far greater amounts of information onto a tiny cluster of atoms. Each "bit" of information, which can be either 1 or 0, now becomes a "qubit," a complex mixture of 1's and 0's with vastly more storage.

The point of bringing up quantum computers is that they may hold the key to exploring the universe. In principle, a quantum computer may give us the ability to exceed human intelligence. They are still a wild card. We don't know when quantum computers will arrive or what their full potential may be. But they could prove invaluable in space exploration. Rather than simply build the settlements and cities of the future, they may take us a step further and give us the ability to do the high-level planning necessary to terraform entire planets.

Quantum computers would be vastly more potent than ordinary digital computers. Digital computers might need several centuries to crack a code based on an exceptionally difficult math problem, such as factorizing a number in the millions into two smaller numbers. But quantum computers, calculating with a high number of mixed atomic states, could swiftly complete the decryption. The CIA and other spy agencies are acutely aware of their promise. Among the mountains of classified material from the National Security Agency that were leaked to the press a few years ago was a top-secret document indicating that quantum computers were being carefully monitored by the agency but that no breakthrough was expected in the immediate future.

Given the excitement and hubbub over quantum computers, when might we expect to have them?

WHY DON'T WE HAVE QUANTUM COMPUTERS?

Computing on individual atoms can be both a blessing and a curse. While atoms can store an enormous quantity of information, the most minute impurity, vibration, or disturbance could ruin a calculation. It is necessary, but notoriously difficult, to totally isolate the atoms from the outside world. They must reach a state of what is called "coherence," in which they vibrate in unison. But the slightest interference—say, someone sneezing in the next building—could cause the atoms to vibrate randomly and independently of one another. "Decoherence" is one of the biggest problems we face in the development of quantum computers.

Because of this problem, quantum computers today can only perform rudimentary calculations. In fact, the world record for a quantum computer involves about twenty qubits. This may not seem so impressive, but it is truly an achievement. It may take several decades or perhaps until late in this century to attain a high-functioning quantum computer, but when the technology arrives, it will dramatically augment the power of AI.

ROBOTS IN THE FAR FUTURE

Considering the primitive state of automatons today, I also would not expect to see self-aware robots for a number of decades—again perhaps not until the end of the century. In the intervening years, we will likely first deploy sophisticated remote-controlled machines to continue the work of exploring space, and then, perhaps, automatons with innovative learning capabilities to begin laying the foundations for human settlements. Later will come self-replicating automatons to complete infrastructure, and then, finally, quantum-fueled conscious machines to help us establish and maintain an intergalactic civilization.

Of course, all this talk of reaching distant stars raises an important question. How are we, or our robots, supposed to get there? How accurate are the starships we see every night on TV?

Why go to the stars?
Because we are the descendants of
those primates who chose to look
over the next hill.
Because we won't survive here
indefinitely.
Because the stars are there, beckoning
with fresh horizons.
—JAMES AND GREGORY BENFORD

8 BUILDING A STARSHIP

In the movie **Passengers**, the **Avalon**, a state-of-the-art starship powered by massive fusion engines, is traveling to Homestead II, a colony on a distant planet. The ads for this settlement are alluring. The Earth is old, tired, overpopulated, and polluted. Why not make a fresh start in an exciting world?

The journey takes 120 years, during which

passengers are placed in suspended animation, their bodies frozen in pods. When the **Avalon** reaches its destination, the ship will automatically awaken its five thousand riders. They will arise from their pods feeling refreshed and ready to build a new life in a new home.

However, during the trip, a meteor storm punctures the ship's hull and damages its fusion engines, causing a cascade of malfunctions. One of the passengers is revived prematurely, with ninety years left to go in the voyage. He becomes lonely and depressed by the thought that the ship will not land until long after he is dead. Desperate for companionship, he decides to wake up a beautiful fellow traveler. Naturally, they fall in love. But when she finds out that he deliberately roused her almost a century too soon, and that she, too, will die in interplanetary purgatory, she goes ballistic.

Movies like **Passengers** embody recent attempts by Hollywood to inject a little realism into its science fiction. The **Avalon** makes its trip the old-fashioned way, never exceeding the speed of light. But ask any kid to imagine a starship, and he or she will come up with something like the **Enterprise** from **Star Trek** or the **Millennium Falcon** from **Star Wars**—capable of whisking crews across the galaxy at a faster-than-light clip, and perhaps even tunneling through space-time and zapping across hyperspace.

Realistically, our first starships may not be manned and may not resemble any of the huge, sleek vehicles dreamed up in films. In fact, they may be no bigger than a postage stamp. In 2016, my colleague Stephen Hawking startled the world by backing a project called Breakthrough Starshot, which seeks to develop "nanoships," sophisticated chips placed on sails energized by a huge bank of powerful laser beams on Earth. The chips would each be the size of your thumb, weigh less than an ounce, and contain billions of transistors. One of the most promising aspects of the endeavor is that we can use existing technology to make it happen instead of having to wait one hundred or two hundred years. Hawking claimed that nanoships could be developed for $10 billion in the span of one generation and, using one hundred billion watts of laser power, would be able to travel at one-fifth the speed of light to reach the Centauri system, the nearest star system, in twenty years. By contrast, remember that each space shuttle mission remained in near-Earth orbit but cost almost $1 billion per launch.

Nanoships would be able to accomplish what chemical rockets never can. Tsiolkovsky's rocket equation shows that it is impossible for a conventional Saturn rocket to reach the nearest star, since it would need exponentially more fuel the faster it went, and a chemical rocket simply cannot carry enough fuel for a journey of such length. Assuming it could reach the nearby stars, the trip would take about seventy thousand years.

Most of the energy of a chemical rocket goes into lifting its own weight into space, but a nanoship passively receives its energy from external ground-based lasers, so there is no wasted fuel—100 percent of it goes into propelling the ship. And since nanoships do not have to generate their own energy, they have no moving

This laser sail, containing a tiny chip as its payload, can be propelled by a beam of lasers to reach 20 percent of the speed of light.

parts. This significantly reduces the chances of mechanical breakdowns. They also have no explosive chemicals and would not blow up on the launchpad or in space.

Computer technology has advanced to the stage where we can pack an entire scientific laboratory into a chip. Nanoships would contain cameras, sensors, chemical kits, and solar cells, all designed to make detailed analyses of faraway planets and radio information back to Earth. Because the cost of computer chips has dropped dramatically, we could send thousands of them to the stars in the hope that a few of them might survive the hazardous journey. (The strategy mimics that of Mother Nature, in which plants scatter thousands of tiny seeds to the winds to boost the odds that some will succeed.)

A nanoship whizzing by the Centauri system at 20 percent of the speed of light would have just a few hours to complete its mission. In that time frame, it would locate Earth-like planets and rapidly photograph and analyze them to determine their surface characteristics, temperatures, and the composition of their atmospheres, in particular looking for the presence of water or oxygen. It would also scan the star system for radio emissions, which might indicate the existence of alien intelligence.

Mark Zuckerberg, founder of Facebook, has

publicly supported Breakthrough Starshot, and Russian investor and former physicist Yuri Milner has personally pledged $100 million. Nanoships are already much more than an idea. But there are several obstacles we must reckon with before we can fully execute the project.

PROBLEMS WITH LASER SAILS

To send a fleet of nanoships to Alpha Centauri, a laser bank would have to fire a barrage of beams totaling at least one hundred gigawatts at the parachutes of the ships for about two minutes. The light pressure from these laser beams would send the ships darting into space. The beams must be aimed with astonishing precision to ensure that the ships hit their target. The slightest deviation in their trajectory would compromise the mission.

The main hurdle we face is not the basic science, which is already available, but funding, even with several high-profile scientists and entrepreneurs on board.

Each nuclear power plant costs several billion dollars and can generate only one gigawatt, or a billion watts, of power. The process of soliciting federal and private financing for a sufficiently powerful and accurate laser bank is causing a severe bottleneck.

As a practice run before aiming for distant stars, scientists may decide to send nanoships to closer destinations within the solar system. It would take them only five seconds to zip to the moon, about an hour and a half to get to Mars, and a few days to reach Pluto. Rather than waiting ten years for a mission to the outer planets, we could receive new information about them from nanoships in a matter of days, and in this way we could observe the developments in the solar system very nearly in real time.

In a subsequent phase of the project, we might attempt to set up a battery of laser cannons on the moon. When a laser passes through the Earth's atmosphere, about 60 percent of its energy is lost. A lunar launch facility would help to remedy this problem, and solar panels on the moon could provide cheap and plentiful electrical energy to fuel the laser beams. Recall that one lunar day is equivalent to about thirty Earth days, so the energy could be efficiently collected and stored in batteries. This system would save us billions of dollars, because unlike nuclear power, sunlight is free.

By the early twenty-second century, the technology for self-replicating robots should be perfected, and we may be able to entrust machines with the task of constructing solar arrays and laser batteries on the moon, Mars, and beyond.

We would ship over an initial team of automatons, some of which would mine the regolith and others of which would build a factory. Another set of robots would oversee the sorting, milling, and smelting of raw materials in the factory to separate and obtain various metals. These purified metals could then be used to assemble laser launch stations—and a new batch of self-replicating robots.

We might eventually have a bustling network of relay stations throughout the solar system, perhaps stretching from the moon all the way to the Oort Cloud. Because the comets in the Oort Cloud extend roughly halfway to Alpha Centauri and are largely stationary, they may be ideal locations for laser banks that could provide an extra boost to nanoships on their journey to our neighboring star system. As each nanoship passed by one of these relay stations, its lasers would fire automatically and give the ship an added push to the stars.

Self-replicating robots could build these distant outposts by using fusion instead of sunlight as the basic source of energy.

LIGHT SAILS

Laser-propelled nanoships are just one type in a much larger category of starships called light sails. Just as sailboats capture the force of the wind, light sails harness the light pressure from sunlight or lasers. In fact, many of the equations used to guide sailboats can also be applied to light sails in outer space.

Light is made up of particles called photons, and when photons strike an object they do exert a minuscule pressure. Because light pressure is so small, scientists were not aware of its existence for a long time. It was Johannes Kepler who first noticed the effect when he realized that, contrary to expectations, comet tails always point away from the sun. Kepler correctly surmised that pressure from sunlight creates these tails by blowing dust and ice crystals in comets away from the sun.

The prescient Jules Verne anticipated light sails in **From the Earth to the Moon** when he wrote, "There will some day appear velocities far greater than these, of which light or electricity will probably be the mechanical agent . . . we shall one day travel to the moon, the planets, and the stars."

Tsiolkovsky further developed the concept of solar sails, or spaceships that utilize light pressure from the sun. But the history of solar sails has been spotty. NASA has not made them a priority. The Planetary Society's Cosmos 1 in 2005 and NASA's NanoSail-D in 2008 both suffered launch failures. They were followed by NASA's NanoSail-D2, which entered low-Earth orbit in 2010. The only successful attempt to send a solar sail past Earth orbit was accomplished by the Japanese in 2010. The IKAROS satellite deployed a sail that was forty-six feet by forty-six feet in size and was powered by solar light pressure. It reached Venus in six months, thereby proving that solar sails were feasible.

The idea continues to percolate despite its erratic progress. The European Space Agency is considering launching the Gossamer solar sail, whose purpose would be to "deorbit" some of the thousands of pieces of space junk littering the area around Earth.

I recently interviewed Geoffrey Landis, an MIT-educated NASA scientist working on the Mars program as well as on light sails. Both he and his wife, Mary Turzillo, are award-winning science fiction novelists. I asked him how he managed to bridge such different worlds—one populated by meticulous scientists and their complex equations, the other filled with space groupies and UFO buffs. He responded that science fiction was wonderful because it allowed him to speculate far into the future. Physics, he said, kept him grounded.

Landis's specialization is the light sail. He has proposed a starship for the journey to Alpha Centauri that would consist of a light sail made of an ultrathin layer of a diamond-like material several hundred miles across. The ship would be gigantic, weighing a million tons, and would require resources from across the solar system to build and operate, including energy from laser banks near Mercury. To be able to stop at its destination, the ship would contain a large "magnetic parachute," with the field produced by a loop of wire sixty miles in diameter. Hydrogen atoms from space would pass through the loop and generate friction, which would gradually slow down the light sail over several decades. A round-trip to Alpha Centauri and back would take two centuries, so the crew would have to be multigenerational. Although this starship is physically achievable, it would be costly, and Landis conceded that it might take fifty to one hundred years to actually assemble and test. In the meantime, he is helping to build the Breakthrough Starshot laser sail.

ION ENGINES

In addition to laser propulsion and solar sails, there are a number of other potential ways to energize a starship. To compare them, it is useful to introduce a concept called "specific impulse," which is the thrust of the rocket multiplied by the time over which the rocket fires. (Specific impulse is measured in units of seconds.) The longer a rocket fires its engines, the larger its specific impulse, from which its final velocity can be calculated.

Here is a simple chart that ranks the specific impulse of several types of rockets. I have not included some designs—like the laser rocket, solar sail, and ramjet fusion rocket—that technically have a specific impulse of infinity, since their engines can be fired indefinitely.

ROCKET ENGINE	SPECIFIC IMPULSE
Solid fuel rocket	250
Liquid fuel rocket	450
Nuclear fission rocket	800 to 1,000
Ion engine	5,000
Plasma engine	1,000 to 30,000
Nuclear fusion rocket	2,500 to 200,000
Nuclear pulsed rocket	10,000 to 1 million
Antimatter rocket	1 million to 10 million

Notice that chemical rockets, which burn for only a few minutes, have the lowest specific impulse. Next on the list are the ion engines, which may be useful for missions to nearby planets. Ion engines start by taking a gas like xenon, stripping the electrons off its atoms to turn them into ions (charged fragments of atoms), and then accelerating these ions with an electric field. The inside of an ion engine bears some resemblance to the inside of a TV monitor, where electric and magnetic fields guide a beam of electrons.

The thrust of ion engines is so excruciatingly

small—often measured in ounces—that when you turn one on in the lab, nothing seems to happen. But once in space, over time they can attain velocities exceeding chemical rockets. Ion engines have been compared to the tortoise in the race with the hare—which, in this case, would be chemical rockets. Although the hare can sprint with enormous speed, it can only do so for a few minutes before it is exhausted. The tortoise, on the other hand, is slower but can walk for days and thus wins long-distance competitions. Ion rockets can operate for years at a time and hence have considerably larger specific impulses than chemical rockets.

To increase the power of an ion engine, one might ionize the gas using microwaves or radio waves and then use magnetic fields to accelerate the ions. This is called a plasma engine, which, in theory, could cut the travel time to Mars from nine months to fewer than forty days, according to its proponents, but the technology is still in development. (One limiting factor to plasma engines is the large amount of electricity necessary to create the plasma, which may even require a nuclear power plant for interplanetary missions.)

NASA has studied and built ion engines for decades. For example, the Deep Space Transport, which may take our astronauts to Mars in the 2030s, uses ion propulsion. Late in this century, ion engines will most likely become the backbone of interplanetary space missions. Although chemical rockets might still be the best option for time-sensitive missions, ion engines would be a solid, dependable choice when time is not the most important consideration.

Beyond the ion engine on the specific impulse chart are propulsion systems that are more speculative. We will discuss each of them in the following pages.

100 YEAR STARSHIP

In 2011, DARPA and NASA funded a symposium entitled the 100 Year Starship. It generated considerable interest. The aim was not to build an actual starship within one hundred years but to assemble top scientific minds who could lay out a feasible agenda for interstellar travel for the next century. The project was organized by members of the Old Guard, an informal group of elderly physicists and engineers, many now in their seventies, who seek to draw upon their collective knowledge to take us to the stars. They have passionately kept the flame alive for decades.

Landis is a member of the Old Guard. But there is also an unusual pair among them, James and Gregory Benford, twins who happen to both be physicists as well as science fiction writers.

James told me that his fascination with starships began when he was a child devouring all the science fiction he could get his hands on, especially Robert Heinlein's old Space Cadet series. He realized that if he and his brother were serious about space, they would have to learn physics. Lots of it. So both set off to get their Ph.D.s in the field. James is now the president of Microwave Sciences and has worked for many decades with high-powered microwave systems. Gregory is a professor of physics at the University of California, Irvine, and in his other life has won the coveted Nebula Award for one of his novels.

In the wake of the 100 Year Starship symposium, James and Gregory wrote a book, **Starship Century: Toward the Grandest Horizon,** containing many of the ideas presented there. James, an expert on microwave radiation, believes that light sails are our best chance of travel beyond the solar system. But, he said, there is a long history of alternate theoretical designs that would be exceedingly expensive but are based on solid physics and might one day actually happen.

NUCLEAR ROCKETS

This history goes back to the 1950s, an era when most people lived in terror of nuclear war but a few atomic scientists were looking for peaceful applications for nuclear energy. They considered all sorts of ideas, such as deploying nuclear weapons to carve out ports and harbors.

Most of these suggestions were rejected due to concerns about the fallout and disruption from nuclear explosions. One intriguing proposal that lingered, however, was called Project Orion, and it sought to use nuclear bombs as the power source for starships.

The skeleton of the plan was simple: create mini atomic bombs and eject them one by one from the back end of a starship. Each time a mini nuke exploded, it would create a shockwave of energy that would push the starship forward. In principle, if a series of mini nukes were released in succession, the rocket could accelerate to nearly the speed of light.

The idea was developed by nuclear physicist Ted Taylor along with Freeman Dyson. Taylor was famous for designing a wide variety of nuclear bombs, from the largest fission bomb ever detonated (with a force of about twenty-five times the Hiroshima bomb) down to the little Davy Crockett portable nuclear canon (with a force one thousand times smaller than the Hiroshima bomb). But he longed to channel his extensive knowledge of nuclear explosives toward peaceful purposes. He jumped at the opportunity to pioneer the Orion starship.

The main challenge was figuring out how to carefully control the sequence of small detonations so that the starship could safely ride the wave of nuclear blasts without being destroyed in the process. Different designs for a range of speeds were drawn up. The largest model would be a quarter of a mile in diameter, would weigh eight million metric tons, and would be propelled by 1,080 bombs. On paper, it could attain a velocity of 10 percent of the speed of light and reach Alpha Centauri in forty years. Despite the immense size of this ship, calculations showed that it might just work.

Critics converged on the idea, however, pointing out that nuclear pulse starships would unleash radioactive fallout. Taylor countered that fallout is created when dirt and the metallic bomb casing become radioactive after the bomb is set off, so it could be avoided if the starship only fired its engine in outer space. But the Test Ban Treaty of 1963 also made it difficult to experiment with miniature atomic bombs. The Orion starship ultimately wound up as a curiosity relegated to old science books.

DRAWBACKS TO NUCLEAR ROCKETS

Another reason the project came to a close was that Ted Taylor himself lost interest. I once asked him why he withdrew his support for the effort, since it seemed like a natural use for his talent. He explained to me that to create the Orion would be to produce a new type of nuclear bomb. Although he spent most of his life designing uranium fission bombs, he realized that one day the Orion spacecraft might use powerful, specially designed H-bombs as well.

These bombs, which release the greatest amount of energy known to science, have gone through three stages of development. The first H-bombs of the 1950s were gigantic devices that required large ships to transport them. For all practical purposes, they would have been useless in a nuclear war. Second-generation nuclear bombs are the small, portable MIRVs, or multiple independently targetable reentry vehicles, that make up the backbone of the U.S. and Russian nuclear arsenals. You can pack ten of them into the nose cone of an intercontinental ballistic missile.

Third-generation nuclear bombs, sometimes called "designer nuclear bombs," are, at the moment, still a concept. They could be easily concealed and custom-made for specific battlefields—for example, the desert, the forest, the Arctic, or outer space. Taylor told me that he had become disillusioned with the project and feared that terrorists could get hold of them. It would be an unspeakable nightmare for him if his bombs fell into the wrong hands and destroyed an American city. He reflected candidly on the irony of his about-face.

He had contributed to a field in which scientists would put pins, each representing a nuclear bomb, in a map of Moscow. But when faced with the possibility that third-generation weapons could put pins in an American city, he suddenly decided to oppose the development of advanced nuclear weapons.

James Benford informed me that although Taylor's nuclear pulse rocket never made it off the drawing board, the government actually did produce a series of nuclear rockets. Instead of exploding mini atomic bombs, these rockets used an old-fashioned uranium reactor to generate the necessary heat. (The reactor was used to heat up a liquid, such as liquid hydrogen, to a high temperature, and then shoot it out a nozzle in the back, creating thrust.) Several versions were built and tested in the desert. These reactors were quite radioactive and there was always the danger of a meltdown during the launch phase, which would have been disastrous. Due to an assortment of technical problems as well as growing anti-nuclear sentiment among the public, these nuclear rockets were mothballed.

FUSION ROCKETS

The scheme to employ nuclear bombs to propel starships died in the 1960s, but in the wings was another possibility. In 1978, the British Interplanetary Society initiated Project Daedalus. Instead of using uranium fission bombs, Daedalus would use mini H-bombs, which Taylor himself looked at but never developed. (The mini H-bombs of Daedalus are actually small second-generation bombs, not the true third-generation bombs that Taylor had so feared.)

There are several ways in which to release the power of fusion peacefully. One process, called magnetic confinement, involves placing hydrogen gas in a large magnetic field the shape of a doughnut and then heating it up to millions of degrees. Hydrogen nuclei smash into one another and are fused into helium nuclei, releasing bursts of nuclear energy. The fusion reactor can be used to heat up a liquid, which is then released through a nozzle, thereby propelling the rocket.

The leading fusion reactor using magnetic confinement at present is called the International Thermonuclear Experimental Reactor (ITER), located in southern France. It is a monstrous machine, ten times bigger than its closest competitor. It weighs 5,110 tons, stands thirty-seven feet tall and sixty-four feet in diameter, and has cost more than $14 billion so far. It is expected to attain fusion by 2035 and ultimately produce five hundred megawatts of heat energy (compared to one thousand megawatts of electricity in a standard uranium nuclear power plant). It is hoped that it will be the first fusion reactor to generate more energy than it consumes. Despite a series of delays and cost overruns, physicists I have talked to are betting that the ITER reactor will make history. We will have our answer before too long. As Nobel laureate Pierre-Gilles de Gennes once said, "We say that we will put the sun into a box. The idea is pretty. The problem is we don't know how to make the box."

Another variation of the Daedalus rocket might be fueled by laser fusion, in which giant laser beams compress a pellet of hydrogen-rich material. This process is called inertial confinement. The National Ignition Facility (NIF), based at the Livermore National Laboratory in California, exemplifies this process. Its battery of laser beams—192 gigantic beams in 4,900-foot-long tubes—is the largest in the world. When the laser beams are focused on a tiny sample of hydrogen-rich lithium deuteride, their energy incinerates the surface of the material, resulting in a mini explosion that causes the pellet to collapse and raises its temperature to one hundred million degrees Celsius. This creates a fusion reaction that unleashes five hundred trillion watts of power in a few trillionths of a second.

I saw a demonstration of the NIF while hosting a Discovery/Science Channel documentary.

Visitors must first pass a series of national security checks, because the U.S. nuclear arsenal is designed at the Livermore Laboratory. When I finally entered, it was overwhelming. A five-story apartment building could easily fit in the main chamber where the laser beams converge.

One version of Project Daedalus exploits a process similar to laser fusion. Instead of a laser beam, it uses a large bank of electron beams to heat the hydrogen-rich pellet. If 250 pellets are detonated per second, enough energy could conceivably be generated for a starship to reach a fraction of the

This image shows the comparative size of the Daedalus fusion starship with the Saturn V rocket. Because of its enormous size, it would most likely have to be assembled in space by robots.

speed of light. However, this design would require a fusion rocket of truly immense size. One version of the Daedalus rocket would weigh fifty-four thousand metric tons and would be about 625 feet long, with a maximum velocity of 12 percent of the speed of light. It is so big it would have to be constructed in outer space.

The nuclear fusion rocket is conceptually sound, but fusion power has not yet been demonstrated. Furthermore, the sheer size and complexity of these projected rockets cast doubt on their feasibility, at least in this century. Still, alongside the light sail, the fusion rocket holds the most promise.

ANTIMATTER STARSHIPS

Fifth wave technologies (which include antimatter engines, light sails, fusion engines, and nanoships) may open up exhilarating new horizons for starship design. Antimatter engines, as in **Star Trek,** may become a reality. They would utilize the greatest energy source in the universe, the direct conversion of matter into energy through matter and antimatter collisions.

Antimatter is the opposite of matter, meaning that it has the opposite charge. An anti-electron has a positive charge, while an anti-proton has a negative charge. (I tried to investigate antimatter in high school by placing a capsule of sodium-22, which emits anti-electrons, in a cloud chamber and photographing the beautiful tracks left by the antimatter. Then I constructed a 2.3-million-electron volt betatron particle accelerator in the hope of analyzing antimatter's properties.)

When matter and antimatter collide, both are annihilated into pure energy, so the reaction releases energy with 100 percent efficiency. A nuclear weapon, by contrast, is only 1 percent efficient; most of the energy inside a hydrogen bomb is wasted.

An antimatter rocket would be rather simple in design. The antimatter would be stored in secure containers and fed into a chamber in steady streams. It would combine explosively with ordinary matter in the chamber and result in a burst of gamma rays and X-rays. The energy would then be shot through an opening in the exhaust chamber to create thrust.

As James Benford remarked to me, antimatter rockets are a favorite concept among science fiction fans, but there are serious problems with building them. For one, antimatter is naturally occurring, but only in relatively small quantities, so we would have to manufacture large amounts of it for use in engines. The first anti-hydrogen atom, with an anti-electron circling around an anti-proton, was created in 1995 at the European

Organization for Nuclear Research (CERN) in Geneva, Switzerland. A beam of ordinary protons was produced and shot through a target made of ordinary matter. That collision resulted in a few particles of anti-protons. Huge magnetic fields separated the protons from the anti-protons by driving them in different directions—one bending to the right, the other to the left. The anti-protons were then slowed down and stored in a magnetic trap, where they were combined with anti-electrons to form anti-hydrogen. In 2016, physicists at CERN took anti-hydrogen and analyzed the anti-electron shells that orbit the anti-proton. As expected, they found an exact correspondence between the energy levels of anti-hydrogen and ordinary hydrogen.

CERN scientists have announced, "If we could assemble all the antimatter we've ever made at CERN and annihilate it with matter, we would have enough energy to light a single electric light bulb for a few minutes." A whole lot more would be needed for a rocket. Also, antimatter is the most expensive form of matter in the world. At today's prices, a gram would go for about $70 trillion. Currently, it can only be created (in very small amounts) with particle accelerators, which are extremely costly to construct and operate. The Large Hadron Collider (LHC) at CERN is the most powerful particle accelerator in the world and cost more than $10 billion to set up, but it can only produce a very thin beam of antimatter. It would bankrupt the United States to accumulate enough to fuel a starship.

The giant atom smashers of today are all-purpose machines, used purely as research tools, and are highly inefficient in their production of antimatter. One partial solution might be to establish factories specifically designed to churn it out. In that case, Harold Gerrish of NASA believes that the cost of antimatter could go down to $5 billion per gram.

Storage presents another difficulty and expense. If you put antimatter in a bottle, sooner or later, it would hit the walls of the bottle and annihilate the container. Penning traps would be needed to enclose it properly. These traps would use magnetic fields to hold atoms of antimatter in suspension and prevent them from coming into contact with the vessel.

In science fiction, issues of cost and storage are sometimes eliminated by the discovery of a deus ex machina—an anti-asteroid that enables us to mine antimatter cheaply. But this hypothetical scenario raises a complicated question: Where does antimatter come from, anyway?

Everywhere we look in outer space with our instruments, we see matter, not antimatter. We know this because the collision of one electron with an anti-electron releases a minimum energy of 1.02 million electron volts. This is the fingerprint of an antimatter collision. But when we examine the universe, we detect very little of this type of radiation. Most of the universe we see around us is made of the same ordinary matter we are made of.

Physicists believe that at the instant of the Big Bang, the universe was in perfect symmetry and there was an equal amount of matter and antimatter. If so, the annihilation between the two would have been perfect and complete, and the universe should be made of pure radiation. Yet here we are, made of matter, which should not be around anymore. Our very existence defies modern physics.

We have not yet figured out why there is more matter than antimatter in the universe. Only one ten-billionth of the original matter in the early universe survived this explosion, and we are part of it. The leading theory is that something violated the perfect symmetry between matter and antimatter at the Big Bang, but we don't know what it is. There is a Nobel Prize waiting for the enterprising individual who can solve this problem.

Antimatter engines are on the short list of priorities for anyone who wants to build a starship. But the properties of antimatter are still almost totally unexplored. It is not known, for example, whether it falls up or down. Modern physics predicts that it should fall down, like ordinary matter. If so, then antigravity would probably not be possible. However, this, along with so much else, has never been tested. Based on cost and our limited understanding, antimatter rockets will probably remain a dream for the next century, unless we happen upon an anti-asteroid drifting in space.

RAMJET FUSION STARSHIPS

The ramjet fusion rocket is another enticing concept. It would look like a giant ice cream cone and would scoop up hydrogen gas in interstellar space, then concentrate it in a fusion reactor to generate energy. Like a jet or a cruise missile, the ramjet rocket would be quite economical. Because jets gulp ordinary air, they do not have to carry their own oxidizer, which reduces cost. Since there is an unlimited amount of hydrogen gas in space for fuel, the spaceship should be able to accelerate forever. As with the solar sail, the engine's specific impulse is infinite.

The famous novel **Tau Zero** by Poul Anderson is about a ramjet fusion rocket that suffers a malfunction and cannot shut down. As it accelerates toward the speed of light, bizarre relativistic distortions begin to occur. Time slows down within the rocket, but the universe around it ages as usual. The faster it goes, the slower time

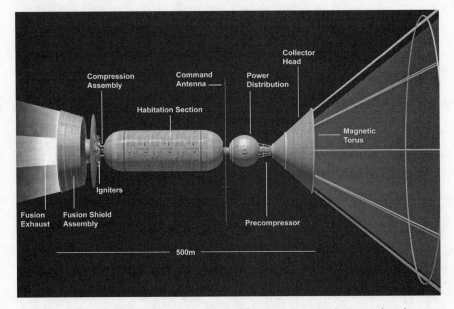

This image shows a ramjet fusion starship, which scoops up hydrogen from interstellar space and burns it in a fusion reactor.

beats inside it. To someone on the starship, however, things seem perfectly normal inside, while the universe outside ages rapidly. Eventually, the starship goes so fast that millions of years pass outside the ship as the crew members watch helplessly. After traveling uncounted billions of years into the future, the crew realizes that the universe is no longer expanding but is actually shrinking. The expansion of the universe is finally reversing. The temperature soars as the galaxies begin to come together toward the final Big Crunch. At the end of the story, just as all the stars are collapsing, the rocket ship manages to skim past the cosmic fireball and witness a Big

Bang as a new universe is born. As fantastic as this tale may be, its foundations do conform to Einstein's theory of relativity.

Apocalyptic narratives aside, the ramjet fusion engine at first might seem too good to be true. But over the years, a number of possible criticisms have been leveled at it. The scoop might have to be hundreds of miles across, which would be both impractically large and prohibitively costly. The rate of fusion might not produce enough energy to sustain a starship. Dr. James Benson also pointed out to me that our sector of the solar system does not contain enough hydrogen to feed the engines, though perhaps other areas of the galaxy might. Others claim that the drag on a ramjet engine as it moves in the solar wind would exceed its thrust, so it could never reach relativistic velocities. Physicists have tried to modify the design to rectify these disadvantages, but we have a long way to go before ramjet rockets become a realistic option.

PROBLEMS WITH STARSHIPS

It should be emphasized that all the starships mentioned so far face other problems associated with traveling near light speed. Asteroid collisions would present a major risk, and even tiny asteroids could pierce the hull of the ship. As we mentioned, the space shuttle suffered small nicks and scars from cosmic debris, which probably hit the spacecraft near orbiting velocity, or eighteen thousand miles per hour. Near light speed, however, impacts will take place at many times that velocity, potentially pulverizing the starship.

In the movies, this hazard is eliminated by powerful force fields that conveniently repel all these micrometeorites—but those unfortunately only exist in the minds of science fiction writers. In reality, electric and magnetic force fields can indeed be generated, but even household objects that are not charged, such as plastic, wood, and plaster, could easily penetrate them. In outer space, tiny micrometeorites, because they are uncharged, cannot be deflected by electric and magnetic fields. And gravitational fields are attractive and extremely weak, so they would not be suitable for the repulsive force fields we would need.

Braking is another challenge. If you're zipping through space at a velocity approaching light speed, how do you slow down when you reach your destination? Solar and laser sails depend on the energy of the sun or banks of laser beams, which cannot be used to decelerate the starship. So they may be useful mainly in flyby missions.

Perhaps the best way to brake nuclear rockets is to turn them around 180 degrees so the thrust is in the opposite direction. However, this strategy would consume roughly half the mission's thrust to reach the targeted velocity and the other half to slow the rocket down. For solar sails, perhaps the sail can be reversed so that light from the star at the destination can be used to slow down the spacecraft.

Another issue is that most of these starships capable of carrying astronauts would be hefty and could only be assembled in outer space. Scores of space missions would be required to send the building materials into orbit, and still more to assemble the pieces. To avoid insurmountable expenses, a more economical method of launching missions into space must be devised. That is where the space elevator may come in.

ELEVATORS INTO SPACE

Space elevators would be a game-changing application of nanotechnology. A space elevator is a long shaft that stretches from the Earth into outer space. You would enter the elevator, press the up button, and then be rapidly lifted into orbit. You wouldn't suffer the crushing g-forces experienced when a booster rocket blasts off its launchpad. Instead, your ride into space would be as mild as taking the elevator to the top of a department store. Like Jack's beanstalk, the space elevator would seemingly defy gravity and provide an effortless way to ascend into the skies.

The possibility of a space elevator was first explored by the Russian physicist Konstantin Tsiolkovsky, who was intrigued by the building of the Eiffel Tower in the 1880s. If engineers could build such a magnificent structure, he asked himself, why not keep going and extend one into outer space? Using simple physics, he was able to show that, in principle, if the tower was long enough, then centrifugal force would be sufficient to keep it upright, without any external force. Just as a ball on a string does not fall to the floor because of its spin, a space elevator would be kept from collapsing by the centrifugal force of the spinning Earth.

The notion that perhaps rockets were not the only way to enter space was radical and exciting. But there was an immediate roadblock. The stress on space elevator cables might reach one hundred gigapascals of tension, which exceeds the breaking point of steel, which is two gigapascals. Steel cables would snap, and the space elevator would come tumbling down.

The concept of space elevators was shelved for almost a hundred years. They were mentioned occasionally by authors like Arthur C. Clarke, who featured them in a novel called **The Fountains of Paradise**. However, asked when a space elevator might be possible, he replied, "Probably about fifty years after everyone stops laughing."

But no one is laughing anymore. Suddenly,

space elevators don't seem so far-fetched after all. In 1999, a preliminary NASA study assessed that an elevator with a cable three feet wide and thirty thousand miles long could transport fifteen tons of payload. In 2013, the International Academy of Astronautics issued a 350-page report projecting that with enough funding and research, a space elevator capable of carrying multiple twenty-ton payloads might be possible by 2035. Price estimates usually range from $10 billion to $50 billion—a fraction of the $150 billion that went into the International Space Station. Meanwhile, space elevators could reduce the cost of putting payloads into space by a factor of twenty.

The problem is no longer one of basic physics but of engineering. Serious calculations are now being made to determine whether space elevator cables could be made of pure carbon nanotubes, which are so strong that they would not break. But can we make enough of these nanotubes to stretch thousands of miles into space? At present, the answer is no. Pure carbon nanotubes are extremely difficult to manufacture beyond a centimeter or so. You might hear announcements that nanotubes many feet long have been constructed, but those materials are actually composites. They consist of tiny threads of pure carbon nanotubes compressed into a fiber and lose the wondrous properties of pure nanotubes.

To stimulate interest in projects like the space

elevator, NASA sponsors the Centennial Challenges program, which awards prizes to amateurs who can invent advanced technologies for the space program. It once held a contest calling for entrants to submit components for a mini-elevator prototype. I participated in it for a TV special I hosted, following a group of young engineers who were convinced that space elevators would open up the heavens to the average person. I watched as they used laser beams to send a small capsule up a long cable. Our TV special tried to capture the enthusiasm of this new class of entrepreneurial engineers, keen to build the future.

Space elevators would revolutionize our access to outer space, which, instead of remaining the exclusive territory of astronauts and military pilots, could become a playground for children and families. They would offer an efficient new approach to space travel and industry and make possible the extraterrestrial assembly of complex machinery, including starships that can travel almost as fast as light.

But realistically, given the enormous engineering problems facing us, a space elevator might not be possible until late in this century.

Of course, considering our restless curiosity and ambition as a species, we will eventually move on beyond fusion and antimatter rockets and face the greatest challenge of all. There is the possibility that one day we might break the ultimate speed limit in the universe: the speed of light.

WARP DRIVE

One day, a boy read a children's book and changed world history. It was 1895, and cities were beginning to be wired up for electricity. To understand this strange new phenomenon, the boy picked up **Popular Books on Natural Science** by Aaron Bernstein. In it, the author asked readers to imagine riding alongside an electric current inside a telegraph wire. The boy then wondered what it would be like if you replaced the electric current with a beam of light. Can you outrace light? He reasoned that since light was a wave, the light beam would look stationary, frozen in time. But even at the age of sixteen, he grasped that no one had ever seen a stationary wave of light. He spent the next ten years puzzling over this question.

Finally, in 1905, he found the answer. His name was Albert Einstein, and his theory was called special relativity. He discovered that you cannot outrace a light beam, because the speed of light is the ultimate velocity in the universe. If you approach it, strange things happen. Your rocket becomes heavier, and time slows down inside it. If you were to somehow reach light speed, you would be infinitely heavy and time would stop. Both conditions are impossible, which means you cannot break the light barrier. Einstein became the cop on the block, setting the ultimate speed limit in the universe. This barrier has bedeviled generations of rocket scientists ever since.

But Einstein was not satisfied. Relativity could explain many of the mysteries of light, but he wanted to apply his theory to gravity as well. In 1915, he came up with an astonishing explanation. He postulated that space and time, which were once thought to be inert and static, were actually dynamic, like smooth bedsheets that can be bent, stretched, or curved. According to his hypothesis, the Earth does not revolve around the sun because it is pulled by the sun's gravity, but because the sun warps the space around it. The fabric of space-time pushes on the Earth so that it moves in a curved path around the sun. Simply put, gravity does not pull. Instead, space pushes.

Shakespeare once said that all the world is a stage and we are actors making our entrances and exits. Picture space-time as an arena. It was once thought to be static, flat, and absolute, with clocks ticking at the same rate across the surface. But in the Einsteinian universe, the stage can be warped. Clocks run at different rates. Actors cannot walk across the stage without falling over. They might claim that an invisible "force" is pulling them in various directions, when actually the warped stage is pushing them.

Einstein also realized that there was a loophole in his general theory of relativity. The larger a star is, the greater the warping of space-time surrounding it. If a star is heavy enough, it becomes a black hole. The fabric of space-time may actually tear, potentially creating a wormhole, which is a gateway or shortcut through space. This concept, first introduced by Einstein and his student Nathan Rosen in 1935, is today called the Einstein-Rosen bridge.

WORMHOLES

The simplest example of an Einstein-Rosen bridge is the looking glass from **Alice's Adventures in Wonderland.** On one side of the looking glass is the countryside of Oxford, England. On the other side is the fantasy world of Wonderland, to which Alice is instantly transported when she puts her finger through the glass.

Wormholes are a favorite plot device in the movies. Han Solo sends the **Millennium Falcon** through hyperspace by propelling it through a wormhole. The refrigerator that Sigourney Weaver's character opens in **Ghostbusters** is a wormhole through which she peers at an entire universe. In C. S. Lewis's **The Lion, the Witch,**

and the Wardrobe, the wardrobe is the wormhole connecting the English countryside to Narnia.

Wormholes were discovered by analyzing the mathematics of black holes, which are collapsed giant stars whose gravity is so intense that even light cannot escape. Their escape velocity is the speed of light. In the past, black holes were thought to be stationary and to have infinite gravity, called a singularity. But all the black holes that have been recorded in space are spinning quite rapidly. In 1963, physicist Roy Kerr discovered that a spinning black hole, if it was moving fast enough, would not necessarily collapse to a pinpoint but to a spinning ring. The ring is stable because centrifugal force prevents it from collapsing. So where does everything that falls into a black hole go? Physicists do not yet know. But one possibility is that matter can emerge from the other side through what is called a white hole. Scientists have looked for white holes, which would release matter rather than swallow it up, but have not found any so far.

If you approached the spinning ring of a black hole, you would witness incredible distortions of space and time. You might see light beams captured billions of years ago by the wormhole's gravity. You might even meet copies of yourself. Your atoms might be stretched by tidal forces in a disturbing and lethal process called spaghettification.

If you entered the ring itself, you might be expelled through a white hole in a parallel universe on the other side. Imagine taking two sheets of paper, held parallel to each other, then drilling a hole through them with a pencil to connect them. If you traveled along the pencil, you would pass between two parallel universes. However, if you passed through the ring a second time, you would arrive at another parallel universe. Each time you went into the ring, you would reach a different universe, in the same way that entering an elevator allows you to move between different floors of an apartment building, except in this case you could never return to the same floor.

Gravity would be finite as you entered the ring, so you would not necessarily be crushed

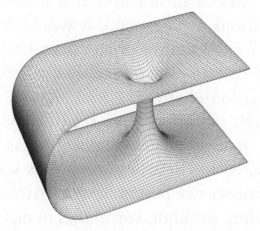

A wormhole is a shortcut that connects two distant points in space and time.

to death. However, if the ring was not spinning fast enough, it could still collapse on you and kill you. But it may be possible to stabilize the ring artificially by adding something called negative matter or negative energy. A stable wormhole is therefore a balancing act, and the key is to maintain the right mixture of positive and negative energy. You need lots of positive energy to naturally create the gateway between universes, as with a black hole. But you also need to create negative matter or energy artificially to keep the gateway open and prevent a collapse.

Negative matter is quite different from antimatter and has never been detected in nature. Negative matter has bizarre antigravitational properties, meaning that it would fall up, rather than down. (By contrast, antimatter is theorized to fall down, not up.) If it existed on the Earth billions of years ago, it would have been repelled by the matter of the planet and would have floated into outer space. Perhaps that's why we haven't found any.

Although physicists have seen no evidence of negative matter, negative energy has actually been created in the laboratory. This keeps alive the hope of science fiction fans who dream of one day traveling through wormholes to distant stars. However, the amount of negative energy that has been created in the laboratory is minuscule, far too small to drive a starship. To create enough negative energy to stabilize a wormhole would require an extremely advanced technology, which we will discuss in more detail in chapter 13. So for the foreseeable future, hyperdrive wormhole starships are beyond our capability.

But recently there has been some excitement generated by another means to warp space-time.

ALCUBIERRE DRIVE

In addition to wormholes, the Alcubierre engine might offer a second way to break the light barrier. I once interviewed the Mexican theoretical physicist Miguel Alcubierre. He was struck with a groundbreaking idea in relativistic physics while watching TV, perhaps the first time this has ever happened. During an episode of **Star Trek,** he marveled that the Starship **Enterprise** could travel faster than light. It could somehow compress the space in front of it so that the stars did not seem as distant. The **Enterprise** did not journey to the stars—the stars came to the **Enterprise.**

Think of moving across a carpet to reach a table. The commonsense way is to walk along the carpet from one point to another. But there is another way. One could rope the table and drag it toward you, so that you are compressing the carpet. So instead of walking across the carpet to reach the table, the carpet folds up and the table comes to you.

An interesting realization dawned on him. Usually, you start with a star or planet and then use Einstein's equations to calculate the bending of space around it. But you can also go backward. You can identify a particular warping and use the same equations to determine the type of star or planet that would cause it. A rough analogy might be made to the way an auto mechanic builds a car. You could begin with the parts that are available—the engine, the tires, and whatnot—and assemble a car from them. Or you could select the design of your dreams and then figure out the parts necessary to create it.

Alcubierre turned Einstein's math on its head, reversing the usual logic of theoretical physicists. He attempted to gauge what kind of star might compress space in the forward direction and expand it in the backward direction. Much to his shock, he reached a very simple answer. It turned out that the space warp used in **Star Trek** was an allowed solution of Einstein's equations! Perhaps warp drive was not so improbable after all.

A starship equipped with Alcubierre drive would have to be surrounded by a warp bubble, a hollow bubble of matter and energy. Space-time inside and outside the bubble would be disconnected. As the starship accelerated, people inside it would feel nothing. They might not think that the ship was moving at all, even though they would be traveling faster than light.

Alcubierre's result shocked the physics community, because it was so novel and radical. But after his paper was published, critics began to point to its weak spots. Although its vision for faster-than-light travel was elegant, it did not address all the complications. If the region inside the starship is separated from the outside world by the bubble, information would not be able to get through, and the pilot would not be able to control the direction of the ship. Steering would be impossible. And then there's the issue of actually creating a warp bubble. In order to compress the space in front of it, it would have to have a certain kind of fuel—that is, negative matter or energy.

We are right back to where we started. Negative matter or negative energy would be the missing ingredient needed to keep our warp bubbles, as well as our wormholes, intact. Stephen Hawking has proven a general theorem stating that **all** solutions of Einstein's equations that allow faster-than-light travel must involve negative matter or energy. (In other words, positive matter and energy that we see in stars can warp space-time so that it perfectly describes the motion of heavenly bodies. But negative matter and energy warp space-time in bizarre ways, creating an antigravi-

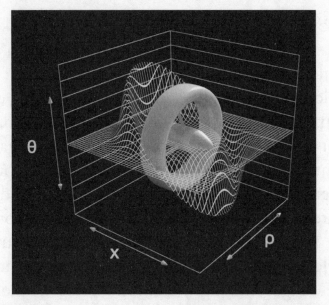

The Alcubierre drive goes faster than light, using Einstein's equations. But it is still controversial whether such a starship can be built.

tational force that can stabilize wormholes and prevent them from collapsing and propel warp bubbles to faster-than-light velocities by compressing space-time in front of them.)

Physicists then tried to calculate the amount of negative matter or energy necessary to propel a starship. The latest results indicate that the amount required is equivalent to the mass of the planet Jupiter. This means that only a very advanced civilization will be able to use negative matter or energy to propel their starships, if it is possible at all. (However, it is possible that the amount of negative matter or energy necessary to go faster than light could drop, because the calculations depend on the geometry and size of the warp bubble or wormhole.)

Star Trek gets around this inconvenient hurdle by postulating that a rare mineral called the dilithium crystal is the essential component of a warp drive engine. Now we know that "dilithium crystals" may be a fancy way of saying "negative matter or energy."

CASIMIR EFFECT AND NEGATIVE ENERGY

Dilithium crystals do not exist, but, tantalizingly, negative energy does, leaving open the possibility of wormholes, compressed space, and even time machines. Although Newton's laws do not allow negative energy, quantum theory does through the Casimir effect, which was proposed in 1948 and measured in the laboratory in 1997.

Say that we have two parallel metal plates that are uncharged. When they are separated by a large distance, we say that there is zero electrical force between them. But as they get closer, they mysteriously begin to attract each other. We can then extract energy from them. Since we start with zero energy but obtain positive energy when the plates are brought together, it follows that the plates themselves originally had negative energy. The reason is rather esoteric. Common sense tells us that a vacuum is a state of emptiness, with zero energy. But actually, it is teeming with matter and antimatter particles that materialize briefly out of the vacuum and then annihilate back into it. These "virtual" particles appear and disappear so rapidly that they do not violate the conservation of matter and energy—that is, the principle that the total amount of matter and energy in the universe always remains the same. This constant churning in the vacuum creates pressure. Since there is more matter and antimatter activity outside the plates than between them, this pressure pushes the plates together, creating negative energy. This is the Casimir effect, which, in quantum theory, demonstrates that negative energy can exist.

Originally, because the Casimir is such a tiny force, it could only be measured with the most sensitive equipment available. But nanotechnology has advanced to the point at which we can tinker with individual atoms. For a TV special I once hosted, I visited a laboratory at Harvard that had a small tabletop device that could manipulate atoms. In the experiment I observed, it was difficult to prevent two atoms that have been brought close to each other from flying apart or coming together due to the Casimir force, which can be either repulsive or attractive. Negative energy may seem like the holy grail to a physicist building a starship, but for a nanotechnologist, the Casimir force is so strong at the atomic level that it becomes a nuisance.

In conclusion, negative energy does exist, and if enough negative energy could somehow be collected, we could, in principle, create a wormhole machine or a warp drive engine, fulfilling some of the wildest fantasies of science fiction. But these technologies are still a long way off, and will be discussed in chapters 13 and 14. In the meantime, we will make do with the light sails that might be zooming through space by the end of the century, offering the first close-up pictures of exoplanets orbiting other stars. By the twenty-second century, we may be able to visit these planets ourselves on fusion rockets. And if we can solve the intricate engineering problems in front of us, we may even be able to make antimatter engines, ramjet engines, and space elevators a reality.

Once we have starships, what will we find in deep space? Will there be other worlds that can sustain humanity? Fortunately, our space telescopes and satellites have given us a detailed look at what lurks among the stars.

Hence I say that I have not merely the opinion, but the strong belief, on the correctness of which I would stake even many of the advantages of life, that there are inhabitants in other worlds.
—IMMANUEL KANT

The desire to know something of our neighbors in the immense depths of space does not spring from idle curiosity nor from thirst for knowledge, but from a deeper cause, and it is a feeling firmly rooted in the heart of every human being capable of thinking at all.
—NIKOLA TESLA

9 KEPLER AND A UNIVERSE OF PLANETS

Every few days, Giordano Bruno has his revenge.

Bruno, Galileo's predecessor, was burned alive at the stake for heresy in Rome in 1600. The stars in the heavens are so numerous, he observed, that our sun must be one of many. Surely these other stars, too, are orbited by a multitude of planets, some of which may even be inhabited by other beings.

The church imprisoned him for seven years without trial, then stripped him naked, paraded him through the streets of Rome, tied his tongue with a leather strap, and lashed him to a wooden pillar. He was given one last chance to recant, but he refused to take back his ideas.

To squelch his legacy, the church placed all his texts on the Index of Forbidden Books. Unlike Galileo's works, Bruno's were banned until 1966. Galileo merely claimed that the sun, not the Earth, was the center of the universe. Bruno suggested that the universe had no center at all. He was one of the first in history to posit that the universe might be infinite, in which case the Earth would be just another pebble in the sky. The church could no longer claim to be the center of the universe, because it had none.

In 1584, Bruno summed up his philosophy, writing, "This space we declare to be infinite . . . in it are an infinity of worlds of the same kind as our own." Now, more than four hundred years later, roughly four thousand extrasolar planets in the Milky Way have been documented, and the list grows almost daily. (In 2017, NASA listed 4,496 candidate planets, of which 2,330 have been confirmed, discovered by the Kepler spacecraft.)

If you go to Rome, you might want to visit the Campo de' Fiori—the "Plain of Flowers"—where there is an imposing statue of Bruno on the very spot where he faced his death. When I went, I found a bustling square full of shoppers, who may not all have been aware that the location had been an execution site for heretics. But Bruno's statue itself gazes down upon a number of young rebels, artists, and street musicians who, unsurprisingly, congregate there. While taking in this peaceful scene, I wondered what kind of atmosphere could have existed back in Bruno's day to inflame such a murderous mob. How could they be whipped up to torture and kill a vagabond philosopher?

Bruno's ideas languished for centuries, because

finding an extrasolar planet is exceedingly difficult and was once thought to be nearly impossible. Planets do not give off light of their own. Even the reflected light of one is about a billion times dimmer than that of the mother star, the harsh glare of which can obscure the planet from view. But thanks to the giant telescopes and space-based detectors we have today, a flood of recent data has proven Bruno to be correct.

IS OUR SOLAR SYSTEM AVERAGE?

In my childhood, I read an astronomy book that changed the way I understood the universe. After describing the planets, the book concluded that our solar system was probably a typical one, echoing the ideas of Bruno. But it also went much further. It speculated that planets in other solar systems moved in almost perfect circles around their sun, like ours. The ones closer to the sun were rocky, while the ones farther out were gas giants. Our sun was the average Joe of stars.

The notion that we live in a quiet, ordinary suburb of the galaxy was simple and comforting.

But boy, were we wrong.

We now realize that we are the oddballs and that the arrangement of our solar system, with its orderly sequence of planets and near-circular orbits, is rare in the Milky Way. As we begin to explore other stars, we are coming across solar systems catalogued in the Extrasolar Planets Encyclopaedia that are radically different from our own. One day, this encyclopedia of planets may contain our future home.

Sara Seager, professor of planetary science at MIT and one of **Time** magazine's twenty-five most influential figures in space exploration, is a key astronomer behind this encyclopedia. I asked her whether she was interested in science as a child. She admitted to me that actually, she was not, but the moon did catch her attention. She was intrigued by the fact that it seemed to follow her whenever her father drove her around. How could something so far away appear to chase after her car?

(The illusion is caused by parallax. We judge distances by moving our heads. Close objects like trees seem to shift the most, while distant entities like the mountains do not change position at all. But objects immediately next to us that are moving with us also don't appear to change position. Our brains therefore confuse remote objects, like the moon, with adjacent ones, like the steering wheel in the car, and make us think that both are moving consistently alongside us. As a result of parallax, many of the UFOs spotted trailing after our cars are actually sightings of the planet Venus.)

Professor Seager's fascination with the heavens blossomed into a lifelong romance. Parents sometimes buy telescopes for their inquisitive children, but she bought her own first telescope with the money she earned from a summer job. She remembers being fifteen and excitedly talking to two of her friends about an exploding star, named Supernova 1987a, that had just been seen in the sky. It had made history as the closest supernova since 1604, and she was planning to go to a party to celebrate the rare event. Her friends, however, were baffled. They did not know what she was talking about.

Professor Seager went on to convert her enthusiasm and sense of wonder about the universe into a bright career in exoplanet science, a discipline that didn't exist two decades ago but that is one of the hottest fields in astronomy today.

METHODS TO FIND EXOPLANETS

It is not easy to see exoplanets directly, so astronomers find them with a variety of indirect strategies. Professor Seager stressed to me that astronomers are confident of their results because they detect exoplanets in multiple ways. One of the most popular is called the transit method. Sometimes, when analyzing the intensity of starlight, you notice that it weakens periodically.

This dimming is a small effect but indicates the presence of a planet that, from the vantage point of Earth, has moved in front of its mother star, thereby absorbing some of its light. Since the path of the planet can therefore be tracked, its orbital parameters can be calculated.

A Jupiter-sized planet would reduce light from a star like our sun by about 1 percent. For an Earth-like planet, the figure is 0.008 percent. This is like the dimming of a car's headlight if a mosquito passes by it. Fortunately, as Professor Seager explained, our instruments are so sensitive and accurate that they can pick up on the slightest changes in luminosity from multiple planets and prove the existence of entire solar systems. However, not all exoplanets move in front of a star. Some have tilted orbits and, therefore, cannot be observed by the transit method.

Another popular approach is the radial velocity, or Doppler, method, in which astronomers look for a star that seems to move back and forth regularly. If there is a large, Jupiter-sized planet orbiting the star, then the star and its Jupiter are actually orbiting each other. Think of a rotating dumbbell. The two weights, representing the mother star and its Jupiter, turn around a common center.

The Jupiter-sized planet is invisible from a distance, but the mother star can clearly be seen moving in a mathematically precise fashion. The Doppler method can be used to calculate its velocity. (For example, if a yellow star moves toward us, the light waves are compressed, like an accordion, so the yellow light turns slightly bluish. If it moves away from us, its light is stretched and turns reddish. The speed of the star can be determined by analyzing how much the light frequency changes as the star moves toward and away from the detector. This is similar to what happens when the police shine a laser beam on your car. The changes in the reflected laser light can be used to measure how fast you are going.)

Careful examination of the mother star over weeks and months also enables scientists to estimate the mass of the planet using Newton's law of gravity. The Doppler method is tedious, but it led to the discovery of the first exoplanet in 1992, which set off a stampede of ambitious astronomers trying to track down the next one. Jupiter-sized planets were the earliest to be observed because giant objects correspond to the largest movements of the mother star.

The transit method and Doppler method are the two main techniques for locating extrasolar planets, but a few others have been introduced recently. One is direct observation, which, as previously mentioned, is difficult to accomplish. However, Professor Seager is excited by NASA's plans to develop space probes that can carefully and precisely obstruct the light from the mother star, which might otherwise overwhelm the planet.

Gravitational lensing may be a promising alternate method, although it only works if there is perfect alignment between the Earth, the exoplanet, and the mother star. We know from Einstein's theory of gravity that light can bend as it moves near a celestial body, because a large mass can alter the fabric of space-time around it. Even if the object is not visible to us, it will change the trajectory of light, just as clear glass does. If a planet moves directly in front of a distant star, the light will be distorted into a ring. This particular pattern is called an Einstein Ring and signals the presence of a substantial mass between the observer and the star.

RESULTS FROM KEPLER

A big breakthrough came with the 2009 launch of the Kepler spacecraft, which was specifically designed to find extrasolar planets by employing the transit method. It was successful beyond the wildest dreams of the astronomical community. Next to the Hubble Space Telescope, the Kepler spacecraft is probably the most productive space satellite of all time. It is a marvel of engineering, weighing 2,300 pounds with a massive 4.6-foot mirror and bristling with the latest high-tech sensors. Because it has to stare at the same spot in the sky for long periods of time in order to get the best data, it does not orbit the Earth but circles the sun instead. From its perch in deep space, which can be one hundred million miles from Earth, it uses a series of gyroscopes to focus on one four-hundredth of the sky, a small patch in the direction of the constellation Cygnus. Inside that tiny field of vision, Kepler has analyzed about two hundred thousand stars and uncovered thousands of extrasolar planets. It has forced scientists to reevaluate our position in the universe.

Instead of locating other solar systems resembling our own, astronomers came across something totally unexpected: planets of all sizes orbiting stars at all distances. "There are planets out there that have no counterpart in our solar system, some of which are in between the size of the Earth and Neptune, or much smaller than Mercury," Professor Seager reflected. "But today, we still haven't found any copies of our solar system." In fact, there have been so many strange results that astronomers don't have enough theories to accommodate them. "The more we find, the less we understand," she confessed. "The whole thing is a mess."

We are at a loss to explain even the most common of these exoplanets. Many of the Jupiter-sized planets, which have been the easiest to find, are not moving in near-circular trajectories as expected but in highly elliptical orbits.

Some Jupiter-sized planets **are** in circular orbits, but they are so close to the mother star that if they were in our solar system, they would be within the orbit of Mercury. These gas giants are called "hot Jupiters," and the solar wind is constantly blowing their atmosphere into outer space. But astronomers once believed that Jupiter-sized planets originate in deep space, billions of miles from the mother star. If so, how did they get so close?

Professor Seager admits that astronomers don't know for sure. But the most probable answer took them by surprise. One theory states that all gas giants form in the outer regions of a solar system, where there is plenty of ice around which hydrogen and helium gas and dust can collect. But in some cases, there is also a large amount of dust spread out within the plane of the solar system. The gas giant may gradually lose energy from the friction of moving through the dust, entering into a death spiral toward the mother star.

This explanation introduced the heretical idea of migrating planets, which had been previously unheard of. (As they edge closer to their suns, they might cross the path of a small Earth-like planet and fling it into outer space. That smaller rocky planet might become a rogue planet, drifting alone in outer space independent of any star. So we don't expect any Earth-like planets in a solar system with Jupiter-sized planets in highly elliptical orbits, or orbits near the mother star.)

In hindsight, these strange results should have been anticipated. Because our own solar system has planets moving in nice circles, astronomers naturally assumed that the balls of dust and hydrogen and helium gas that become solar systems condensed evenly. We now realize that it is more likely that gravity compresses them in a haphazard, random way, resulting in planets that move in elliptical or irregular orbits that may intersect or collide with one another. This is important because it may be that only solar systems with circular planetary orbits like ours are conducive to life.

EARTH-SIZED PLANETS

Earth-like planets are small and hence cause faint dimming or subtle distortions of light from the mother sun. But with the Kepler spacecraft and giant telescopes, astronomers have begun to locate "super-Earths," which, like Earth, are rocky and capable of sustaining life as we know it but are 50 percent to 100 percent larger than our planet. We cannot yet account for their origin, but in 2016 and 2017, a series of sensational, headline-grabbing discoveries about them were made.

Proxima Centauri is the closest star after our sun to the Earth. It is actually part of a triple star system and orbits a pair of larger stars called Alpha Centauri A and B, which orbit each other. Astronomers were stunned to come across a planet just 30 percent larger than the Earth moving around Proxima Centauri. They named it Proxima Centauri b.

"This is a game changer in exoplanetary science," declared Rory Barnes, an astronomer at the University of Washington in Seattle. "The fact that it's so close means we have the opportunity to follow up on it better than any other planet discovered so far." The next batch of giant telescopes in development, like the James Webb Space Telescope, might be able to capture the first photograph of the planet. As Professor Seager put it, "It's absolutely phenomenal. Who would have thought that after all these years of wondering about planets that there's one around our nearest star?"

Proxima Centauri b's mother star is a dim red dwarf only 12 percent as massive as the sun, so the planet must be relatively close to the star in order to be inside its habitable zone, where it can support liquid water and possibly even oceans. The radius of the planet's orbit is just 5 percent of the radius of the Earth's orbit around the sun. It also revolves around its mother star much faster, making one complete revolution every 11.2 days. There is intense speculation about whether Proxima Centauri b has conditions compatible with life as we know it. One major concern is that the planet might be bombarded by solar winds, which could be two thousand times more intense than those hitting the Earth. To shield itself against these blasts, Proxima Centauri b would have to have a strong magnetic field. At present, we do not have enough information to determine whether this is the case.

It has also been suggested that Proxima Centauri b may be tidally locked, so that, like our own moon, one side always faces the star. That side would be perpetually hot, while the other side would be permanently cold. Liquid water oceans might then occur only at the narrow band between these two hemispheres, where the temperature is moderate. However, if the planet has a dense-enough atmosphere, the winds might equalize the temperatures so that liquid oceans could exist freely across its surface.

The next step is to determine the composition

of the atmosphere and whether it contains water or oxygen. Proxima Centauri b was detected using the Doppler method, but the chemical composition of its atmosphere is best assessed with the transit method. When an exoplanet crosses directly in front of the mother star, a tiny sliver of light passes through its atmosphere. Molecules of certain substances in the atmosphere absorb specific wavelengths of starlight, allowing scientists to determine the nature of those molecules. However, for this to work, the orientation of the exoplanet's path must be just right, and there is only a 1.5 percent chance that Proxima Centauri b's orbit is aligned correctly.

It would be an astonishing coup to find molecules of water vapor on an Earth-like planet. Professor Seager explained that "if you have a small rocky planet, you can only have water vapor if you also have liquid water on its surface. So if we find water vapor on a rocky planet, we can infer that it also has liquid oceans."

SEVEN EARTH-SIZED PLANETS AROUND ONE STAR

Another unprecedented finding came in 2017. Astronomers located a solar system that violated all the theories of planetary evolution. It contained seven Earth-sized planets orbiting a mother star called TRAPPIST-1. Three of the planets are in the Goldilocks zone and may have oceans. "This is an amazing planetary system, not only because we have found so many planets, but because they are all similar in size to the Earth," said Michaël Gillon, the leader of the Belgian scientific group that made the discovery. (The name **TRAPPIST** is both an acronym for the telescope used by the group and a reference to Belgium's popular beer.)

TRAPPIST-1 is a red dwarf a mere thirty-eight light-years from Earth, and its mass is only 8 percent that of the sun. Like Proxima Centauri, it has a habitable zone. If transposed over our own solar system, the orbits of all seven planets would fit inside Mercury's path. The planets take less than three weeks to circle the mother star, and the innermost makes a complete revolution in thirty-six hours. Because the solar system is so compact, the planets interact gravitationally and could, in theory, disrupt their own arrangement and collide. One might naïvely expect them to careen into one another. But fortunately, an analysis in 2017 showed that they are in resonance, meaning that their orbits are in phase with one another and no collisions will take place. The solar system seems to be stable. But as with Proxima Centauri b, astronomers are investigating the possible effects of solar flares and tidal locking.

On **Star Trek,** whenever the **Enterprise** is about to encounter an Earth-like planet, Spock announces that they are approaching a "class M planet." Actually, there's no such thing in astronomy—yet. Now that thousands of different types of planets have made their debut, including a variety of Earth-like planets, it's only a matter of time before a new nomenclature is introduced.

TWIN OF THE EARTH?

If a planetary twin of the Earth exists in space, it has eluded us so far. But we have found about fifty super-Earths so far. Kepler-452b, which was discovered by the Kepler spacecraft in 2015 and is about 1,400 light-years from us, is particularly interesting. It is 50 percent bigger than our planet, so you would weigh more than you do on the planet Earth, but otherwise, living there may not be so different from living on Earth. Unlike the exoplanets that orbit around a red dwarf, it circles a star that is only 3.7 percent more massive than the sun. Its period of revolution is 385 Earth days, and its equilibrium temperature is 17 degrees Fahrenheit, slightly warmer than the Earth. It lies within the habitable zone. Astronomers searching for extraterrestrial intelligence trained their radio telescopes to receive messages from any civilization that might be on the planet but have detected none as yet. Unfortunately, because Kepler-452b is so far away, even the next generation of telescopes will not be able to collect significant information about its atmospheric composition.

Kepler-22b, which is six hundred light-years away and 2.4 times the size of the Earth, is also being studied. Its orbit is 15 percent smaller than Earth's—it completes one revolution in 290 days—but the luminosity of its mother star, Kepler-22, is 25 percent lower than the sun's. These two effects compensate for each other, so the surface temperature of the planet is believed to be comparable to that of the Earth. It also lies in the habitable zone.

But KOI 7711 is the exoplanet that is getting the most attention because, as of 2017, it is the one with the most Earth-like features. It is 30 percent larger than Earth, and its mother star is very much like our own. It is not at risk of being fried by solar flares. The length of one year on the planet is almost identical to a year on Earth. It is in the habitable zone of its star, but we do not yet have the technology to evaluate whether its atmosphere contains water vapor. All conditions seem right for it to host some form of life. However, at 1,700 light-years away, it is the farthest exoplanet of the three.

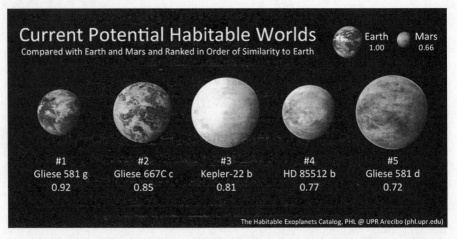

Current Potential Habitable Worlds
Compared with Earth and Mars and Ranked in Order of Similarity to Earth

Earth 1.00   Mars 0.66

#1	#2	#3	#4	#5
Gliese 581 g	Gliese 667C c	Kepler-22 b	HD 85512 b	Gliese 581 d
0.92	0.85	0.81	0.77	0.72

The Habitable Exoplanets Catalog, PHL @ UPR Arecibo (phl.upr.edu)

This illustration shows the relative size of the Earth compared to super-Earths that have been discovered orbiting other stars.

After analyzing scores of these planets, astronomers have discovered that they can usually be arranged into two categories. The first is the super-Earths (like those in the image on the previous page) we have been discussing. "Mini Neptunes" is the other. They are gaseous planets about two to four times the size of the Earth and do not resemble anything in our immediate vicinity; our Neptune is four times bigger than Earth. Once a small planet is discovered, astronomers try to determine which category it belongs to. This is like biologists trying to classify a new animal as either being a mammal or reptile. One mystery is why these categories aren't represented in our own solar system when they seem to be so prominent elsewhere in space.

ROGUE PLANETS

Rogue planets are among the strangest celestial bodies that have been discovered so far. They wander the galaxy without orbiting any particular star. They probably originated in a solar system but got too close to a Jupiter-sized exoplanet and were hurled into deep space. As we have seen, these large Jupiter-sized planets frequently have elliptical orbits or migrate in a spiral toward the mother star. It is likely that their paths intersected with smaller planets, and as a consequence, rogue planets might be more plentiful than ordinary ones. In fact, according to some computer models, our own solar system may have ejected ten or so rogue planets billions of years ago.

Because rogue planets are not near a light source and give off no light themselves, it seemed hopeless at first to try to locate them. But astronomers have been able to find some through the gravitational lensing technique, which requires a very precise and quite rare alignment to take place between the background star, the rogue planet, and the detector on Earth. As a result, one has to scan millions of stars in order to detect a handful of rogues. Fortunately, this process can be automated so that computers, not astronomers, do the searching.

Thus far, 20 potential rogue planets have been

identified, one of which is only seven light-years from Earth. However, another recent study, conducted by Japanese astronomers who examined fifty million stars, found even more possible candidates, up to 470 rogue planets. They estimated that there might be 2 rogue planets for every star in the Milky Way. Other astronomers have speculated that the number of rogue planets could exceed the number of ordinary ones by a factor of one hundred thousand.

Can life as we know it exist on rogue planets? It depends. Like Jupiter or Saturn, some may have a large number of ice-covered moons. If so, tidal forces could melt the ice into oceans, where life may originate. But in addition to sunlight and tidal forces, there is a third way in which a rogue planet may have an energy source that could give birth to life: radioactivity.

An episode from the history of science might help to illustrate this point. In the late nineteenth century, a simple calculation done by the physicist Lord Kelvin showed that Earth should have cooled down a few million years after its creation and therefore should be frozen solid and inhospitable to life. This result sparked a debate with biologists and geologists, who insisted that the Earth was billions of years old. The physicists were shown to be wrong when Madame Curie and others discovered radioactivity. It is the nuclear force at the core of the Earth, from long-lived radioactive elements like uranium, that has kept Earth's core hot for billions of years.

Astronomers have conjectured that rogue planets, too, might have radioactive cores that keep them relatively warm. This means a radioactive core could supply heat to hot springs and volcanic vents on the bottom of an ocean where the chemicals of life may be created. So if rogue planets are as numerous as some astronomers believe, then the most probable place to find life in the galaxy may not be within the habitable zone of a star but on the rogue planets and their moons.

ODDBALL PLANETS

Astronomers are also researching a plethora of completely startling planets, some of which defy categorization.

In the movie **Star Wars,** the planet Tatooine revolves around two stars. Some scientists scoffed at this idea, because such a planet would be in an unstable orbit and would collapse into one of the stars. But planets circling three stars have been documented, as in the Centauri system. We've even found four-star systems, in which two sets of double stars move around each other.

Another planet has been discovered that apparently may be made of diamonds. It is called

55 Cancri e and is about double the size of the Earth but weighs about eight times more. In 2016, the Hubble Space Telescope successfully analyzed its atmosphere—the first time this had ever been done with a rocky exoplanet. It detected hydrogen and helium but no water vapor. Later, the planet was found to be rich in carbon, which might constitute about a third of its mass. It is also scalding hot, with a temperature of 5,400 degrees Kelvin. One theory postulates that the heat and pressure in the core may be extraordinary enough to give rise to a diamond planet. However, these glittering deposits, if they indeed exist, are forty light-years from us, so mining them is beyond our current capabilities.

Possible water worlds and ice worlds have also been located. This is not necessarily unanticipated. It is believed that our own planet, early in its history, was covered in ice—a Snowball Earth. At other times, when the Ice Ages receded, the planet was flooded with water. Gliese 1214 b, the first of six known potentially water-covered exoplanets to be identified, was found in 2009. It is forty-two light-years away and six times larger than the Earth. It lies outside the habitable zone, orbiting seventy times closer to its mother sun than the Earth does. It may get as hot as 280 degrees Celsius, so life as we know it probably cannot exist. But by using various filters to analyze light scattered through the planet's atmosphere as it transits the mother star, significant amounts of water have been confirmed. The water may not be in familiar liquid form due to the planet's temperature and pressure. Instead, Gliese 1214 b might be a steam planet.

We have come to a striking realization about the stars, as well. We once thought that our yellow star was typical in the universe, but astronomers now believe that dim red dwarf stars, which emit only a fraction of the light of our sun and usually cannot be seen with the naked eye, are the most common. By one estimate, 85 percent of the stars in the Milky Way are red dwarfs. The smaller a star is, the slower it burns hydrogen fuel and the longer it can shine. Red dwarfs may last for trillions of years, far longer than the ten-billion-year life span of our sun. Perhaps it is not surprising that Proxima Centauri b and the TRAPPIST system both involve red dwarfs, because they are so numerous. Thus the area around these stars may be one of the most promising sites to search for more Earth-like planets.

CENSUS OF THE GALAXY

The Kepler spacecraft has surveyed enough planets in the Milky Way galaxy that a rough census can be made. The data indicate that, on average, every star you see has some kind of planet orbiting around it. About 20 percent of the stars, like our sun, have Earth-like planets—that is, are similar to the Earth in size and are in the habitable zone. Since there are roughly one hundred billion stars in the Milky Way, about twenty billion Earth-like planets may exist in our backyard. In fact, this is a conservative estimate—the actual number could be much higher.

Unfortunately, the Kepler spacecraft, after sending a mountain of information that changed the way we conceptualize the universe, began to malfunction. One of its gyroscopes started to fail in 2013, and it lost the ability to lock onto planets.

But further missions are being planned that will continue to augment our understanding of exoplanets. In 2018, the Transiting Exoplanet Survey Satellite (TESS) will be launched. Unlike the Kepler, it will scan the entire sky. TESS will examine two hundred thousand stars over a two-year period, concentrating on stars that are thirty to one hundred times brighter than those inspected by the Kepler, including all the possible Earth-sized planets or super-Earths in our region of the galaxy, a number astronomers expect to be around five hundred. Furthermore, the James Webb Space Telescope, the replacement for the Hubble Space Telescope, will be inaugurated shortly and should be able to actually photograph some of these exoplanets.

Earth-like planets may be prime targets for future starships. Now that we are on the cusp of investigating them in depth, it is important to explore two considerations: living in outer space, with the biological demands it would entail, and encountering life in space. We must first take a look at our existence on the Earth and how it may be enhanced to meet new challenges. We may have to modify ourselves, extending our life span, adjusting our physiology, and even altering our genetic heritage. We will also have to contend with the possibility of discovering anything from microbes to advanced civilizations on these planets. Who might be out there, and what would it mean for us to meet them?

PART III LIFE IN THE UNIVERSE

The aeons involved in traversing the galaxy are not daunting to immortal beings.
—SIR MARTIN REES, ASTRONOMER ROYAL OF ENGLAND

10 IMMORTALITY

The movie **The Age of Adaline** is the tale of a woman born in 1908 who is caught in a snowstorm and freezes to death. Fortunately, she is hit by a freak lightning bolt, which revives her. This peculiar event changes her DNA, and she mysteriously stops aging.

As a result, she remains young while her friends and lovers grow old. Inevitably, suspicions and rumors start, and she is forced to leave town. Instead of rejoicing in her limitless youth, she withdraws from society and rarely speaks to anyone. Immortality, instead of being a gift, is a curse to her.

Finally, she is hit by a car and dies in the accident. In the ambulance, the electric shock from the defibrillator not only revives her, it reverses the genetic effects of the lightning bolt, and she becomes mortal. Instead of weeping at her loss of immortality, she rejoices when she finds her first gray hair.

While Adaline eventually rejects the promise of immortality, science is actually moving in the other direction, making enormous strides in understanding aging. Scientists concerned with deep space exploration are keenly interested in this research, because the distance between stars is so great that it may take centuries for a ship to complete its voyage. Thus, the process of building a starship, surviving the voyage to the stars, and settling on distant planets might require several lifetimes. In order to survive the journey we would have to build multigenerational ships, put our astronauts and pioneers in suspended animation, or extend their life spans.

Let us explore each of these ways in which humans might travel to the stars.

MULTIGENERATIONAL SHIPS

Assume that an Earth-like twin has been discovered in space that has an oxygen/nitrogen atmosphere, liquid water, a rocky core, and is a size that closely matches that of the Earth. It sounds like an ideal candidate for habitation. But then you realize that this twin is one hundred light-years from Earth. This means that a starship, using perhaps fusion or antimatter propulsion, would require two hundred years to reach it.

If one generation corresponds to roughly twenty years, this means that ten generations of humans will be born on the starship, which will be the only home they know.

Although this may seem daunting, realize that during the Middle Ages, master architects would design grand cathedrals knowing that they would not live long enough to see the completion of their masterpieces. They knew that perhaps their grandchildren would be the ones to celebrate the opening of the cathedral.

Also, realize that during the Great Diaspora, when humans began to leave Africa roughly seventy-five thousand years ago in search of a new home, they realized that perhaps it would take many generations for them to complete their journey.

So the concept of a multigeneration voyage is not a new one.

But there are problems that have to be faced if you are traveling on a starship. First, the population has to be chosen very carefully, with at least two hundred people per ship, in order to have a sustainable breeding population. The number of people has to be monitored so that the population remains relatively constant and they do not exhaust supplies. Even the slightest deviation in population, extended over ten generations, could lead to a disastrous overpopulation or underpopulation, which would threaten the entire mission. So a variety of methods—such as cloning, artificial insemination, and test tube babies—might be required to keep the population stable over time.

Second, resources would have to be carefully monitored as well. Food and waste would have to be recycled constantly. Nothing could be thrown away.

There is also the problem of boredom. For example, people living on small islands often complain of "island fever," the intense feeling of claustrophobia and the burning desire to leave the island and explore new worlds. One possible solution would be to use virtual reality to create imaginary, fanciful worlds, using advanced computer simulations. Another possibility is to create goals, contests, tasks, and jobs for people so that their lives have direction and purpose.

In addition, decisions have to be made on board the ship, such as the allocation of resources and duties. A democratically elected body will have to be created to supervise the day-to-day operation of the ship. But this leaves open the possibility that a future generation might not want to fulfill the original mission or that a charismatic demagogue might take over and subvert it.

There is one way, however, of eliminating many of these problems: resorting to suspended animation.

MODERN SCIENCE AND AGING

In the movie **2001**, a crew of astronauts are kept frozen in pods as their giant ship makes the arduous journey to Jupiter. Their bodily functions are reduced to zero, so there are none of the complications associated with multigenerational starships. Since the passengers are frozen, the mission designers would not have to worry about the astronauts consuming large quantities of resources or keeping the population stable.

But is this really possible?

Anyone who has lived in the north during wintertime knows that fish and frogs can be frozen solid in the ice, and that when spring comes and the ice melts, they will emerge as if nothing happened.

Normally, we would expect the freezing process to kill these animals. As you lower the temperature of blood, ice crystals begin to grow and expand both within the cells, eventually rupturing the cell wall, and also outside the cells, potentially squeezing and crushing them. Mother Nature solves this problem using a simple solution: antifreeze. During winter, we put antifreeze in our cars in order to lower the freezing point of water. In the same way, Mother Nature uses glucose as an antifreeze, thereby lowering the freezing point of blood. So although the animal is frozen in a block of ice, the blood in its veins is still liquid and can still perform basic bodily functions.

For humans this high concentration of glucose in our bodies would be toxic and would kill us. So scientists have experimented with other kinds of chemical antifreezes in a process they call vitrification, which involves using a combination of chemicals to lower the freezing point so that ice crystals do not form. Although it sounds intriguing, the results have been disappointing so far. Vitrification often has adverse side effects. The chemicals used in the labs are often poisonous and can be lethal. To date, no one has ever been frozen solid, then thawed out, and lived to tell about it. So we are a long way from effectively achieving suspended animation. (This hasn't stopped entrepreneurs from prematurely advertising this as a way to cheat death. They claim that people with fatal illnesses can have their bodies frozen, for a hefty fee, and then be revived decades later, when their diseases can be cured. However, there is absolutely no experimental proof that this process works.) Scientists hope that in time, these technical questions might be solved.

So on paper, suspended animation may be the ideal way to solve many of the problems of long-term voyages. Although it is not a practical option today, in the future it might be one of the chief methods of surviving interstellar missions.

However, there is one problem with suspended animation. If there is an unexpected emergency, such as an asteroid impact, then humans may be required to fix the damage. Robots may be activated to make the initial repairs, but, if the emergency is severe enough, human experience and judgment will be required. This might mean that some of the passengers who are engineers may have to be revived, but this last-minute option could be fatal if it takes too long to revive the engineers and human intervention is required immediately. This is the weak point in an interstellar voyage using suspended animation. It may be that a small multigenerational society of engineers would have to be kept awake and ready during the entire voyage.

SEND IN THE CLONES

Yet another proposal to colonize the galaxy is to send embryos containing our DNA into space, in the hopes that one day they may be revived at some distant destination. Or we could send the DNA code itself, to eventually be used to create new humans. (This was the method mentioned in the movie **Man of Steel.** Although Superman's home planet, Krypton, had exploded, the Kryptonians were advanced enough to sequence the DNA of the entire Kryptonian population before the planet blew up. The plan was that this information could be sent to a planet like Earth, and then the DNA sequence could be used to create clones of the original Kryptonians. The only problem was that this might involve taking over the Earth and getting rid of humans, who are unfortunately in the way.)

There are advantages to the cloning approach. Instead of having gigantic starships containing huge artificial Earth-like environments and life support systems, this would only involve transplanting DNA. Even large tanks of human embryos could easily fit inside a standard rocket ship. Not surprisingly, science fiction writers have imagined that this happened aeons ago, when some prehuman species might have spread their DNA in our sector of the galaxy, making possible the rise of humanity.

There are, however, several drawbacks to this proposal. At present, no human has ever been cloned. In fact, no primate has ever been successfully cloned either. The technology is not yet advanced enough to create human clones, although it might be accomplished in the future. If so, robots might be designed to create and take care of these clones.

More important, reviving human clones might create creatures genetically identical to us, but they won't have our memories or personality. They will be a blank slate. At present, the ability to send a person's entire memory and personality in this way is far beyond our capabilities. Again, this requires a technology that will take decades or centuries to create, if it's possible at all.

But rather than being frozen or cloned, perhaps another way one can journey to the stars is to slow down or even stop the aging process.

SEARCH FOR IMMORTALITY

The search for eternal life is one of the oldest themes in all of human literature. It goes back to the **Epic of Gilgamesh,** written nearly five thousand years ago. The poem chronicles the exploits of a Sumerian warrior on a noble quest. Along the way, he has many adventures and encounters, including one with a Noah-like individual who witnessed the Great Flood. The goal of this long journey is to find the secret of immortality. In the Bible, God banished Adam and Eve from the Garden of Eden after they disobeyed Him and ate from the tree of knowledge. God was angry because they might use this knowledge to become immortal.

Humanity has been obsessed with immortality for ages. For much of human history infants died in childbirth, and the lucky ones who survived often lived in a state of near starvation. Epidemics would spread like wildfire because people often threw their kitchen garbage out the window. Sanitation as we know it did not exist, so villages and cities reeked. Hospitals, if they existed at all, were places for the poor to die. They were warehouses for destitute, poverty-stricken patients, since the rich could afford to have private physicians. But the rich were also victims of disease, and their private doctors were little more than quacks. (One midwestern doctor kept a diary of his daily visits to patients. He confessed that there were only two items in his black bag that actually worked. Everything else was snake oil. What actually worked was the hacksaw to cut off injured and diseased limbs, and morphine to dull the pain of amputation.)

In 1900, the official life expectancy in the United States was forty-nine. But two revolutions added decades to that number. First, sanitation improved, which gave us clean water and waste removal and helped to eliminate some of the worst epidemics and plagues, adding about fifteen years to our life expectancy.

The next revolution was in medicine. We often take for granted that our ancestors lived in mortal fear of a bestiary of ancient diseases (like tuberculosis, smallpox, measles, polio, whooping cough, and so on). In the postwar era these diseases were largely conquered by antibiotics and vaccines, adding another ten years to our life expectancy. During this time, the reputation of hospitals changed significantly. They became places where real cures for diseases were dispensed.

So can modern science now unlock the secrets of the aging process, slowing down or even stopping the clock, increasing life expectancy to an almost limitless degree?

This is an ancient quest, but what is new is that it has now gained the attention of some of the richest people on the planet. In fact, there is an influx of Silicon Valley entrepreneurs investing millions to defeat the aging process. Not content to wire up the world, their next goal is to live forever. Sergey Brin, the cofounder of Google, hopes to do nothing less than "cure death." And Calico, led by Brin, may eventually dump billions into a partnership with the pharmaceutical company AbbVie to tackle the problem. Larry

Ellison, cofounder of Oracle, thinks that accepting mortality is "incomprehensible." Peter Thiel, the cofounder of PayPal, wants to live to a modest 120 years, while Russian internet tycoon Dmitry Itskov wants to live to 10,000 years. With the support of people like Brin and technological innovations, we may finally be able to use the full force of modern science to unlock this age-old mystery and extend our life span.

Recently, scientists have revealed some of the deepest secrets of the aging process. After centuries of false starts, there are now a few reliable, testable theories that seem promising. These involve caloric restriction, telomerase, and age genes.

Of these, one and only one method has proven to extend the life span of animals, sometimes even doubling it. It is called caloric restriction, or severely limiting the intake of calories in an animal's diet.

On average, animals that eat 30 percent fewer calories live 30 percent longer. This has been amply demonstrated with yeast cells, worms, insects, mice and rats, dogs and cats, and now primates. In fact, it is the only method that is universally accepted by scientists to alter the life span of all animals that have been tested so far. (The only important animal that has not yet been tested is humans.)

The theory is that animals in the wild naturally live in a state of near starvation. These animals use their limited resources to reproduce during times of plenty, but during hard times they enter a state of near hibernation to conserve their resources and live past the famine. Feeding animals less food triggers the second biological response and they live longer.

One problem with caloric restriction, however, is that these animals become lethargic, sluggish, and lose interest in sex. And most humans would balk at the prospect of eating 30 percent fewer calories. So the pharmaceutical industry would like to find the chemicals that govern this process and harness the power of caloric restriction without its glaring side effects.

Recently, a promising chemical called resveratrol has been isolated. Resveratrol, found in red wine, helps to activate the sirtuin molecule, which has been shown to slow down the oxidation process, a principle component in aging, and therefore it may help protect the body from age-related molecular damage.

I once interviewed Leonard P. Guarente, the MIT researcher who was one of the first to show the link between these chemicals and the aging process. He was surprised at the number of food faddists who have latched onto it as a fountain of youth. He doubted this was the case but left open the possibility that if the true cure for aging was ever found, resveratrol and these chemicals may play some role. He even cofounded a company, Elysium Health, to explore these possibilities.

Another clue to the cause of aging might be telomerase, which helps to regulate our biological clock. Every time a cell divides, the tips of the chromosomes, called telomeres, get a bit shorter. Eventually, after approximately fifty to sixty divisions, the telomeres become so short that they disappear and the chromosome begins to fall apart, so the cell enters a state of senescence and no longer functions correctly. Thus there is a limit to how many times a cell can divide, called the Hayflick limit. (I once interviewed Dr. Leonard Hayflick, who laughed when asked if the Hayflick limit can somehow be reversed to give us the cure for death. He was extremely skeptical. He realized that this biological limit was fundamental to the aging process, but its consequences are still being studied, and because aging is a complex biochemical process involving many different pathways, we are a long way from being able to alter that limit in humans.)

Nobel laureate Elizabeth Blackburn is more optimistic and says, "Every sign, including genetics, says there's some causality [between telomeres] and the nasty things that happen with aging." She notes that there is a direct link between shortened telomeres and certain diseases.

For example, if you have shortened telomeres—if your telomeres are in the bottom third of the population in terms of length—then your risk of cardiovascular disease is 40 percent greater. "Telomere shortening," she concludes, "seems to underlie the risks for the diseases that kill you . . . heart disease, diabetes, cancer, even Alzheimer's."

Recently, scientists have been experimenting with telomerase, the enzyme discovered by Blackburn and her colleagues that prevents the telomeres from shortening. It can, in some sense, "stop the clock." When bathed in telomerase, skin cells can divide indefinitely, far beyond the Hayflick limit. I once interviewed Dr. Michael D. West, then of the Geron Corporation, who experiments with telomerase and claims that he can "immortalize" a skin cell in the lab so that it lives indefinitely. (This has added a new verb to the English language: "to immortalize.") The skin cells in his lab can divide hundreds of times, not just fifty or sixty.

But it should be pointed out that telomerase has to be regulated very carefully, because cancer cells are also immortal and they use telomerase to attain that immortality. In fact, one of the things that separates cancer cells from normal ones is that they live forever and reproduce without limit, eventually creating the tumors that can kill you. So cancer may be an unwanted byproduct of using telomerase.

GENETICS OF AGING

Yet another possibility for defeating aging is through gene manipulation.

The fact that aging is heavily influenced by our genes is readily apparent. Butterflies, after they emerge from the cocoon, live only for a few days or weeks. Mice studied in laboratories usually live only about two years or so. Dogs age about seven times faster than humans and live a little more than ten years.

In looking at the animal kingdom, we also find animals that live so long that their life span is difficult to measure. In 2016, in the journal **Science,** researchers reported that the Greenland shark has an average life span of 272 years, surpassing the 200-year life span of the bowhead whale, making it the longest-lived vertebrate. They calculated the age of these sharks by analyzing the layers of tissue in their eyes, which grow with time, layer for layer, like an onion. They even found one shark that was 392 years old and another that might be as old as 512.

So different species with different genetic makeup vary widely in life expectancy, but even among humans, with our almost identical genes, studies have consistently shown that twins and close relatives have similar life expectancies and that people chosen at random vary far more widely.

So if aging is at least partially governed by genes, the key is to isolate those genes that control it. There are several avenues of approach.

A promising one is to analyze the genes of young people and then compare these genes with those of the old. By comparing the two sets using a computer, one can rapidly isolate where most of the genetic damage caused by aging takes place.

For example, aging in a car takes place mainly in the engine, where the oxidation and wear and tear take the greatest toll. The "engines" of a cell are the mitochondria. That's where sugars are oxidized to extract energy. A careful analysis of the DNA within the mitochondria indicates that errors are indeed concentrated here. The hope is that one day scientists might use the cells' own repair mechanisms to reverse the buildup of errors in the mitochondria and therefore prolong the cells' useful life.

Thomas Perls of Boston University analyzed the genes of centenarians, on the assumption that some people are genetically disposed to live longer, and identified 281 markers for genes that seem to slow down the aging process and somehow make these centenarians less vulnerable to disease.

The mechanism of aging is slowly being revealed, and many scientists are cautiously optimistic that it might be controllable sometime in the coming decades. Their research shows that aging, apparently, is nothing but the accumulation of errors in our DNA and our cells, and perhaps one day we can arrest or even reverse this damage. (In fact, some Harvard professors are so optimistic about their research that they have even set up companies in hopes of capitalizing on the advanced aging research being done in their labs.)

So the fact that our genes play an important role in how long we live is indisputable. The problem arises in identifying which genes are involved in this process, separating out environmental effects, and altering these genes.

CONTROVERSIAL AGING THEORIES

One of the oldest of myths concerning aging is that you can achieve eternal youth by drinking the blood or consuming the soul of the young, as if youth can be transferred from one person to another, as in the vampire legend. The succubus is a beautiful mythical creature who remains eternally youthful because when she kisses you, she sucks the youth from your body.

Modern research indicates there might be a

kernel of truth to this idea. In 1956, Clive M. McCay of Cornell University sewed the blood vessels of two rats together, one old and decrepit and the other young and vigorous. He was astonished to find that the old mouse started to look younger, while the reverse happened to the young mouse.

Decades later, in 2014, Amy Wagers at Harvard University reexamined this experiment. Much to her surprise, she found the same rejuvenation effect among mice. She then isolated a protein called GDF11 that seems to underlie this process. These results were so remarkable that **Science** magazine chose it as one of the ten breakthroughs of the year. But in the years since this astonishing claim, other groups have tried to duplicate this research, with mixed results. It remains unclear whether GDF11 will be a valuable weapon in the quest to fight aging.

Another controversy involves the human growth hormone (HGH), which has created an enormous fad, but its effectiveness in preventing aging is based on very few reliable studies. In 2017, a major study on more than eight hundred subjects by the University of Haifa in Israel found evidence of the opposite effect, that HGH might actually decrease a person's life expectancy. Furthermore, another study indicates that a genetic mutation that results in a reduced

HGH level may lengthen the human life span, so the effect of HGH may backfire.

These studies teach us a lesson. In the past, wild claims concerning aging often faded when analyzed carefully, but today researchers demand that all results be testable, reproducible, and falsifiable, the hallmark of true science.

Biogerontology, a new science that seeks to find the secret of the aging process, is being born. Recently, there has been an explosion of activity in this area, and a host of promising genes, proteins, processes, and chemicals are being analyzed, including FOXO3, DNA methylation, mTOR, insulin growth factor, Ras2, acarbose, metformin, alpha-estradiol, et cetera. Each has generated enormous interest among scientists, but results are still preliminary. Time will tell which avenue promises the best results.

Today, the quest for the fountain of youth, a field once populated by mystics, charlatans, and quacks, is now being tackled by the world's leading scientists. Although a cure for aging does not yet exist, scientists are pursuing many promising avenues of research. Already, they can extend the life span of certain animals, but it remains to be seen if this can be transferred to humans.

Although the pace of research has been incredible, we are still a long way from being able to solve the mystery of aging. Eventually, a way might be found to slow down and even stop the aging process using a combination of several of these avenues. Perhaps the next generation will make the necessary breakthroughs. As Gerald Sussman once lamented, "I don't think the time is quite right, but it's close. I'm afraid, unfortunately, that I'm in the last generation to die."

ANOTHER PERSPECTIVE ON IMMORTALITY

Adaline may have regretted the gift of immortality, and she's probably not alone, but many people still want to stop the effects of aging. A trip to the local pharmacy reveals row after row of over-the-counter potions that claim to reverse the aging process. Unfortunately, all of them are the byproduct of the overheated imagination of Madison Avenue marketers trying to sell snake oil to gullible customers. (According to many dermatologists, the one ingredient in all these "anti-aging" potions that actually works is moisturizer.)

I once hosted a BBC TV special in which I went to Central Park and interviewed some random bystanders. I asked, "If I had the fountain of youth in my hand, would you drink from it?" Surprisingly, every person I interviewed said no. Many said it was a normal thing to age and die.

It was the way it should be, and dying was part of living. I then went to a nursing home, where many of the patients were suffering from the pain and discomforts of aging. Many were beginning to show signs of Alzheimer's and were forgetting who and where they were. When I asked them if they would drink from the fountain of youth, they all eagerly said, "Yes!"

OVERPOPULATION

What happens if we solve the problem of aging? When and if this happens, then the vast distance to the stars may not seem so daunting. Immortal beings may view interstellar travel in a completely different way than we do. They may view the enormous time required to build starships and send them to the stars as just a small obstacle. In the same way we save up months for a long-awaited vacation, immortal beings may view the centuries necessary to visit the stars as nothing more than an annoyance.

It should be pointed out that the gift of immortality may have an unintended consequence, which is to create a vast overpopulation of the Earth. This may place huge strains on the resources, food, and energy of the planet, leading eventually to blackouts, mass migrations, food riots, and conflicts between nations. So immortality, instead of ushering in an Age of Aquarius, may spark a new wave of world wars.

All this, in turn, may help to accelerate the mass exodus from the Earth, providing a safe haven for pioneers who are tired of an overpopulated and polluted planet. Like Adaline, people might realize that the gift of immortality was actually a curse.

But just how serious is this concern with overpopulation? Will it threaten our very existence?

For most of history, the human population was well under 300 million, but with the coming of the industrial revolution, the world population slowly grew to 1.5 billion by 1900. It is now 7.5 billion and grows by a billion every twelve years or so. The UN estimates that by 2100, it will soar to 11.2 billion. Eventually, we may exceed the carrying capacity of the planet, which could mean food riots and chaos, as Thomas Robert Malthus predicted back in 1798.

In fact, overpopulation is one reason why some people advocate going to the stars. But a closer examination of the issue shows that the growth of the world population, although still rising, is slowing down. The U.N., for example, has revised its predictions downward several times. Many demographers, in fact, predict that the world population will begin to level off and might even stabilize late in the twenty-first century.

To understand all these demographic changes, we have to understand the worldview of a peasant. A farmer in a poor country does a simple calculus: every child makes him richer. Children work in the fields and cost very little to raise. Room and board on a farm are almost free. But when you move to the city, the calculus flips the other way. Every child makes you poorer. Your child goes to school, not the fields. Your child has to be fed from the grocery store, which is expensive. Your kid has to live in an apartment, which costs money. So a peasant, once he becomes more urban, wants two kids, not ten. And when the peasant enters the middle class, he wants to enjoy life a bit and may only have one child.

Even in countries like Bangladesh, which does not have much of an urbanized middle class, the birthrate is slowly falling. This is due to the education of women. Studies of numerous nations have found a distinct pattern: the birthrate falls dramatically as a nation industrializes, urbanizes, and educates young girls.

Other demographers have argued that it is a tale of two worlds. On the one hand, we see a continuing rise in the birthrate in poor countries with low levels of education and a weak economy. On the other hand, we see a leveling off of the birthrate and even contraction in some countries as they develop industry and become more prosperous. In any event, an exploding world population, although still a threat, is not as inevitable or terrifying as previously thought.

Some analysts are concerned that we will shortly exceed our food supply. Others, however, have argued that the food problem is actually an energy problem. If one has enough energy, one can increase productivity and crop production to keep up with demand.

On a number of occasions, I have had the opportunity to interview Lester Brown, one of the world's leading environmentalists and founder of the famed Worldwatch Institute, a think tank for the Earth. His organization closely monitors the world's food supply and the state of the planet. He is worried about another factor: Do we have enough food to feed the people of the world as they become middle class consumers? The hundreds of millions of people in China and India who are now entering the middle class watch Western movies and want to emulate that lifestyle, with its wasteful use of resources, large consumption of meat, big houses, fixation on luxury goods, et cetera. He is concerned we may not have enough resources to feed the population as a whole, and certainly would have difficulty feeding those who want to consume a Western diet.

His hope is that as poor nations industrialize,

they do not follow the historic path of the West and instead adopt strict environmental laws to conserve resources. Time will tell if the nations of the world can meet this challenge.

So we see that advances in slowing down or stopping the aging process could have a profound effect on space travel. They could create beings who do not see the vast distances to the stars as an obstacle. They may want to embark upon challenges that take many years, such as building and then sailing starships on voyages that may take centuries.

Furthermore, attempts to alter the aging process may exacerbate overpopulation of the Earth, which in turn may accelerate the exodus from the Earth. Colonists to the stars may be pushed to leave the Earth if overpopulation becomes unbearable.

However, it is still too early to tell which of these trends will dominate the next century. But given the rate at which we are now unraveling the aging process, these developments may come sooner than expected.

DIGITAL IMMORTALITY

In addition to biological immortality, there is a second type, called digital immortality, which raises some interesting philosophical questions.

In the long run, digital immortality may be the most efficient way to explore the stars. If our fragile biological bodies cannot stand the strain of interstellar travel, there is the possibility of sending our consciousness to the stars instead.

When we try to reconstruct our genealogy, we often encounter a problem. After about three generations, the trail runs cold. The vast majority of our ancestors lived and died without leaving any evidence of their existence other than their offspring.

But today, we leave a huge digital footprint. For example, just by analyzing your credit card transactions, it is possible to tell the countries you visit, the food you like to eat, the clothes you wear, the schools you attended. To this, add your blog posts, diaries, emails, videos, photos, et cetera. With all of this information, it is possible to create a holographic image of you that talks and acts just like you, with your mannerisms and memories.

One day, we might have a Library of Souls. Instead of reading a book on Winston Churchill, we might have a conversation with him. We would talk to a projection with his facial gestures, body movements, and voice inflections. The digital record would have access to his biographical data, his writings, and his opinions on political, religious, and personal matters. In all ways it would feel like talking to the man himself. I would personally enjoy having a conversation with Albert Einstein to discuss the theory of relativity. One day your great-great-great-grandchildren may have a conversation with you. This is one form of digital immortality.

But is this really "you"? It is a machine or simulation that has your mannerisms and biographical details. The soul, some would argue, cannot be reduced to information.

But what happens when we are able to reproduce your brain, neuron for neuron, so that all your memories and feelings are recorded? The next level of digital immortality beyond the Library of Souls is the Human Connectome Project, an ambitious effort to digitize the entire human brain.

As Daniel Hillis, cofounder of Thinking Machines, once said, "I'm as fond of my body as anyone, but if I can be 200 with a body of silicon, I'll take it."

TWO WAYS TO DIGITIZE THE MIND

There are actually two separate approaches to digitizing the human brain. The first is the Human Brain Project, in which the Swiss are trying to create a computer program that can simulate all of the brain's basic features using transistors instead of neurons. So far, they have been able to simulate the "thinking process" of a mouse and rabbit for several minutes. The goal of the project is to create a computer that can talk rationally like a normal human being. Its director, Henry Markram, says, "If we build it correctly, it should speak and have an intelligence and behave very much as a human does."

So this approach is electronic—it attempts to duplicate the intelligence of the brain with a vast array of transistors with tremendous computing power. But a parallel approach is being pursued in the United States that is biological instead, trying to map out the neural pathways of the brain.

This approach is called the BRAIN Initiative (Brain Research through Advancing Innovative Neurotechnologies). Its goal is to unravel the neural structure of the brain itself, cell by cell, and ultimately to map the pathways of every neuron in the brain. Since the human brain contains roughly one hundred billion neurons, each connected to about ten thousand other neurons, it at first seems hopeless to create a road map of every neuron. (Even the relatively simple task of mapping the brain of the mosquito involves producing data that can completely fill a room full of CDs from top to bottom.) But computers and robots have radically reduced the time and effort necessary to complete this tedious, herculean task.

One approach is the "slice and dice approach,"

which involves slicing up the brain into thousands of slides and then using microscopes to reconstruct the connections between all the neurons. A much faster approach has recently been proposed by the scientists at Stanford University, who have pioneered a technique called optogenetics. This method involves first isolating a protein called opsin, which is involved in eyesight. When you shine a light on this gene within a neuron, it causes the neuron to fire.

Using genetic engineering, one can implant the gene for opsin into neurons that you want to study. By shining a light on a section of a mouse's brain, a researcher can cause the neurons involved with a certain muscle activity to fire, and the mouse begins to exhibit a specific activity, such as running around. In this way, one can see the precise neural pathways used to control certain types of behaviors.

For example, this ambitious project may help to unravel the secret of mental illness, which is one of the most debilitating of all human maladies. By mapping the human brain, we might be able to isolate the origin of this affliction. (For example, all of us talk to ourselves silently. When we do, the left brain, which controls language, consults the prefrontal cortex. But in schizophrenics, we now know, the left brain activates without permission from the prefrontal cortex, which is the conscious part of the brain. Since the left brain does not talk to the prefrontal cortex, the schizophrenic thinks the voices in his or her head are real.)

Even with these revolutionary new techniques it still may take several more decades of hard work before scientists have a detailed map of the human brain. But when this is finally achieved, perhaps late in the twenty-first century, would this mean that we can upload consciousness into a computer and send it to the stars?

IS THE SOUL JUST INFORMATION?

If we die and our connectome lives on, then are we in some sense immortal? If our mind can be digitized, then is the soul just information? If we can put all the neural circuits and memories of the brain onto a disk and then upload it into a supercomputer, will the uploaded brain function and act as the real brain? Will it be indistinguishable from the real thing?

Some people find this idea repulsive, because if you upload your mind into a computer, then you will spend an eternity trapped inside a sterile machine. Some think it is a fate worse than dying. There was one episode of **Star Trek** featuring a superadvanced civilization where the pure consciousness of an alien was kept inside a glowing sphere. Aeons ago, the aliens had given up their physical bodies and lived inside these spheres ever since. The aliens became immortal, but one of these aliens longed to have a body once again, to be able to feel real sensations and passions, even if it meant forcibly taking over someone else's body.

Although living inside a computer may sound unappealing to some, there is no reason you couldn't have all the sensation of a living, breathing human being. Although your connectome would reside inside a mainframe computer, it could control a robot that looks identical to you. You feel everything the robot experiences, so that, for all intents and purposes, you have the sensation that you are living inside a real body, potentially even one with superpowers. Everything that the robot sees and feels is relayed back to the mainframe computer and incorporated into your consciousness. So controlling the robot avatar from the mainframe is indistinguishable from your actually being "inside" the avatar.

In this way, you could explore distant planets. Your superhuman avatar will be able to withstand blistering temperatures on sun-scorched planets or the freezing temperatures on distant icy moons. A starship carrying the mainframe that houses your connectome could be sent to a new solar system. As the starship reaches a suitable planet, your avatar could be sent down to explore it, even if the planet has a poisonous atmosphere.

An even more advanced form of uploading your mind into a computer was envisioned by computer scientist Hans Moravec. When I interviewed him, he claimed that his method of uploading the human mind could even be done without losing consciousness.

First you would be placed on a hospital gurney, next to a robot. Then a surgeon would take individual neurons from your brain and create a duplicate of these neurons (made of transistors) inside the robot. A cable would connect these transistorized neurons to your brain. As time goes by, more and more neurons are removed from your brain and duplicated in the robot. Because your brain is connected to the robot brain, you are fully conscious even as more and more neurons are replaced by transistors. Eventually, without losing consciousness, your entire brain and all its neurons are replaced by transistors. Once all one hundred billion neurons have been duplicated, the connection between you and the artificial brain is finally cut. When you gaze back at the stretcher, you see your old body, lacking its brain, while your consciousness now exists inside a robot.

But the question remains, Is that really "you"? To most scientists, if a robot can duplicate all of your behavior down to the very last gesture, with all your memories and habits intact, and is indistinguishable from the original person in every way, then they would say that it is "you" for all intents and purposes.

As we've seen the distances between stars are so great that it will take several lifetimes to reach even those closest in our galactic neighborhood. So multigenerational travel, life extension, and the search for immortality may all play an essential role in the exploration of our universe.

Beyond the question of immortality lies a larger question: How far should we extend not just our life span, but our human body? Even more possibilities exist if we alter our genetic heritage. Given the rapid advances in BCI (brain-computer interface) and genetic engineering, it may be possible to create enhanced bodies with new skills and potentials. One day, we might enter the "posthuman" era, and this might be the best way to explore the universe.

> [Aliens might have] capabilities
> indistinguishable from telekinesis,
> ESP, and immortality . . . they may
> have powers that seem magical . . .
> they will be spiritually advanced
> creatures. Perhaps they will have
> solved the riddle of the quantum, and
> will be able to walk through walls. Um,
> gee, they sound sort of like angels.
> —DAVID GRINSPOON

11 TRANSHUMANISM AND TECHNOLOGY

In the movie **Iron Man,** suave industrialist Tony Stark dons a sleek computerized suit of armor, bristling with missiles, bullets, flares, and explosives. It quickly transforms a frail human into a powerful superhero. But the real magic is on the inside of the suit, which is crammed with the latest computer technology, all controlled by a direct connection with Tony Stark's brain. At the speed of thought, he can rocket into the sky or launch his incredible array of weaponry.

As fanciful as **Iron Man** may be, it is now possible to build a version of this device.

This is not just an academic exercise, since one day we may have to alter and enhance our bodies using cybernetics or even change our genetic makeup in order to survive in hostile exoplanetary environments. Transhumanism, instead of being a branch of science fiction or a fringe movement, may become an essential part of our very existence.

Furthermore, as robots become increasingly powerful and even surpass us in intelligence, we may have to merge with them—or face being replaced by our creations.

Let us explore these various possibilities, especially as they relate to exploring and colonizing the universe.

SUPERSTRENGTH

The world was shocked in 1995 when Christopher Reeve, the handsome actor who played Superman in the movies, was tragically paralyzed from the neck down in an accident. Reeve, who soared into space on the screen, was confined permanently to a wheelchair and was able to breathe only with the help of a respirator. His dream was to use modern technology to regain control of his limbs. He died in 2004, just a decade before this was accomplished.

At the 2014 World Cup in São Paolo, Brazil, a man kicked a soccer ball to start the games, an event witnessed by a billion people. This by itself was not remarkable. What was remarkable was that this man was paralyzed. Professor Miguel Nicolelis of Duke University had inserted a chip into the man's brain. The chip was connected to a portable computer that controlled his exoskeleton. By simply thinking, this paralyzed individual was able to walk and kick the ball.

When I interviewed Dr. Nicolelis, he said that when he was a child, he was mesmerized by the Apollo mission to the moon. His goal was to create another sensation, like the moon landing. Wiring that paralyzed patient so he could kick the ball at the World Cup was a dream come true. It was his moon shot.

I once interviewed John Donoghue of Brown University, one of the pioneers of this approach. He told me that it takes a bit of training, like riding a bicycle, but soon his patients are able to control the motion of an exoskeleton and can do simple tasks (such as grabbing a cup of water, operating household appliances, controlling a wheelchair, and surfing the web). This is possible because a computer is able to recognize certain brain patterns associated with specific body movements. The computer can then activate the exoskeleton so that these electrical impulses are converted into action. One of his paralyzed patients was elated that she could grab a cup of soda and drink from it, something that was previously beyond her capability.

Work done at Duke, Brown, Johns Hopkins, and other universities has given the gift of mobility to people who had long given up hope that they would ever move again. And the U.S. military has devoted more than $150 million in a program called Revolutionary Prosthetics to sponsor these devices for the benefit of veterans from Iraq and Afghanistan, many of them suffering from spinal cord injuries. Eventually, thousands of people who are confined to wheelchairs and beds—whether as the result of warfare, car accidents, disease, or sports injuries—may be able to have the use of their limbs back again.

Besides exoskeletons, another possibility is to strengthen the human body biologically to live on a planet with greater gravity. This possibility was raised when scientists discovered a gene that causes muscles to expand. This was first found in mice, when a genetic mutation caused mice to become muscle-bound. The press dubbed it the "Mighty Mouse gene." Later, the human form of this gene was found and was dubbed the "Schwarzenegger gene."

The scientists who isolated this gene expected phone calls from doctors wishing to help their patients suffering from degenerative muscle diseases. They were surprised, however, when half the calls that came in were from bodybuilders who wanted to bulk up. And most of them did not care that this research was experimental, with unknown side effects. Already it is causing headaches for the sports industry, because it is much harder to detect than other forms of chemical enhancement.

Having the ability to control one's muscle mass may prove important if we explore planets that have a gravitational field larger than the Earth's. So far, astronomers have found a large number of super-Earths (rocky planets within the habitable zone that might even have oceans). They seem likely candidates for human habitation, except their gravitational field can be 50 percent greater than Earth's. This means that it might be necessary to increase our muscles and bones in order to thrive on them.

ENHANCING OURSELVES

In addition to enhancing our muscles, scientists have begun to use this technology to sharpen our senses. People who suffer from certain kinds of deafness now have the option of using cochlear implants. These devices are remarkable, capable of transforming the sound waves coming into the ear into electrical signals that can be sent to the auditory nerve and then the brain. Already about half a million people have chosen to have these sensors implanted.

And for some of those who are blind, an artificial retina can restore a limited amount of vision. This device can be located either in an external camera or it can be placed directly on the retina. This device translates visual images into electrical impulses that the brain can then translate back into visual imagery.

One example, the Argus II, consists of a tiny video camera placed in a person's glasses. The images are then sent to an artificial retina, which relays the signals to the optic nerve. This device can create images of about 60 pixels, and an improved version now being tested has a resolution of 240 pixels. (By contrast, the human eye can recognize the equivalent of about a million pixels, and a person needs at least 600 pixels to identify faces and familiar objects.) A German company is experimenting with another artificial retina with 1,500 pixels that, if successful, might allow a vision-impaired person to function almost normally.

Blind individuals who have tried these artificial retinas have been amazed that they can see colors and outlines of images. It is only a matter of time before we have artificial retinas that can rival human eyesight. And beyond that, it may be possible for an artificial retina to see "colors" corresponding to things that are invisible to the human eye. For example, people are often burned in the kitchen because a hot metal pot looks identical to a cold one. This is because our eyes are incapable of seeing infrared heat radiation. But artificial retinas and goggles can be constructed that can easily detect it, such as night-vision goggles used by the military. So, with artificial retinas, a person may have the ability to see that heat signature and also see other forms of radiation that are invisible to us. This super-vision, in turn, may prove invaluable on other planets. On distant worlds conditions will be radically different. The atmosphere may be dark, hazy, or obscured by dust or impurities. It might be possible to create artificial retinas that can "see" through a Martian dust storm via infrared heat detectors. On distant moons, where sunlight is almost nonexistent, these artificial retinas could intensify whatever reflected light there is.

Another example would be a device that detects ultraviolet radiation, which is harmful and can cause skin cancer but is common throughout the universe. On the Earth, we are protected from intense UV light from the sun by our atmosphere, but on Mars, the UV light is unfiltered. Because UV light is invisible, we are often unaware when we are exposed to harmful levels. But someone with super-vision on Mars could immediately see if the UV light is harmful. On a planet like Venus, which is perpetually clouded over, these artificial retinas would be able to use UV light to navigate around the terrain (in the same way that bees detect UV light from the sun to find their way on an overcast day).

Another application of super-vision would be telescopic and microscopic vision. Tiny, special lenses would enable us to see distant objects, or minuscule objects and cells, without having to lug around bulky telescopes and microscopes.

This type of technology may also give us the power of telepathy and telekinesis. Already, it is possible to create a chip that can pick up our brain waves, decipher some of them, and then transmit this information to the internet. For example, my colleague Stephen Hawking, who suffers from ALS, has lost all motor functions, including the movement of his fingers. Today, a chip has been placed in his glasses that can pick up brain waves that are then sent to a laptop and a computer. In this way, he can type messages mentally, albeit slowly.

From there it is a short step to telekinesis (the ability to move objects with the mind). Using this same technology, one can connect the brain directly to a robot or other mechanical device that can then execute our mental commands. It's easy to imagine that, in the future, telepathy and telekinesis will be the norm; we will interact with machines by sheer thought. Our mind will be able to turn on the lights, activate the internet, dictate letters, play video games, communicate with friends, call for a car, purchase merchandise, conjure any movie—all just by thinking. Astronauts of the future may use the power of their minds to pilot their spaceships or explore distant planets. Cities may rise from the deserts of Mars, all due to master builders who mentally control the work of robots.

Of course, this process of enhancing ourselves is not new but has been happening for all of human existence. Throughout history, we see examples of how humans have used artificial means to enhance our power and influence: clothing, tattoos, makeup, headdresses, ceremonial robes, feathers, glasses, hearing aids, microphones, headphones, et cetera. In fact, it seems to be a universal feature of all human societies that we try to tinker with our bodies, especially to increase our chances of reproductive success. The difference, however, between enhancements of the past and the future is that, as we explore the universe, enhancements may be the key to surviving in different environments. In the future, we might live in the mental age, where our thoughts control the world around us.

THE POWER OF THE MIND

Another milestone in brain research was achieved when scientists, for the first time in history, were able to record a memory. Scientists at Wake Forest and the University of Southern California put electrodes on the hippocampus of mice, where short-term memories are processed. They recorded the impulses within the hippocampus as the mice executed simple tasks, such as learning to drink water from a tube. Later, after the mice had forgotten this task, their hippocampus was stimulated by the recording, and the mice remembered immediately. Primate memories have also been recorded with similar results.

The next target may be to record the memories of patients suffering from Alzheimer's disease. Then we can place a "brain pacemaker" or "memory chip" on their hippocampus, which will flood it with memories of who they are, where they live, and who their relatives are. The military has taken a serious interest in this. In 2017, the Pentagon announced a $65 million grant to develop a tiny, advanced chip that can analyze a million human neurons as the brain communicates with a computer and forms memories.

We will need to study and refine this technique, but by the late twenty-first century, it is conceivable that we might be able to upload complex memories into our brain. In principle, we might be able to transfer skills and abilities, even entire college courses, into our brain, enhancing our capabilities almost without limit.

This may prove useful for astronauts of the future. When landing on a new planet or moon, there are so many details to learn and memorize about the new environment and so many technologies that have to be mastered. So uploading memories might be the most efficient way to learn entirely new information about distant worlds.

But Dr. Nicolelis wants to go much farther with this technology. He told me that all these breakthroughs in neurology will eventually give rise to the "brain net," which is the next stage in the evolution of the internet. Instead of transmitting bits of information, brain net will transmit entire emotions, feelings, sensations, and memories.

This could help to break down barriers between people. Often, it is hard to understand other people's point of view, their suffering and anguish. But with brain net, we would be able to experience firsthand the anxieties and fears that trouble others.

This could revolutionize the entertainment industry, in the same way that the talkies rapidly replaced silent movies. In the future, audiences may be able to feel the emotions of the actors, to experience their pain, joy, or suffering. The movies of today may soon be obsolete.

So astronauts of the future may be able to use brain net in important ways. They will be able to mentally communicate with other settlers, instantly exchange vital information, and amuse themselves with an entirely new form of entertainment. Also, since space exploration is potentially dangerous, they will be able to sense a person's mental condition much more accurately than before. When embarking on a new space mission to explore dangerous terrain, having brain net will help astronauts bond and also reveal mental problems, such as depression or anxiety.

There is also the possibility of using genetic engineering to enhance the mind. At Princeton University, a gene was found in mice (dubbed the "smart mouse gene") that increased their ability to navigate mazes. The gene is called **NR2B**, and it is involved in communication between the cells of the hippocampus. Researchers found that when mice lacked the **NR2B** gene, their memory was impaired as they navigated the maze. However, if they had extra copies of the **NR2B** gene, their memory was enhanced.

These researchers placed mice in a shallow pan of water that had an underwater platform they could stand on. Once they found the platform, the smart mice were able to remember instantly where it was and swim directly to it when reintroduced to the environment. Ordinary mice, by contrast, could not remember the location of the platform and swam randomly. So memory enhancement is a possibility.

THE FUTURE OF FLIGHT

Humans have always dreamed of flying like the birds. The god Mercury had tiny wings on his hat and ankles that allowed him to fly. There is also the myth of Icarus, who used wax to attach feathers to his arms in order to fly. Unfortunately, he flew too close to the sun. The wax melted, and he plunged into the ocean. But the technology of the future will finally give us the gift of flight.

On a planet with a thin atmosphere and rugged terrain like Mars, perhaps the most convenient way to travel is the jet pack, a staple of science fiction cartoons and movies. It appeared in the very first Buck Rogers strip back in 1929, when Buck meets his future girlfriend while she is soaring through the air using a jet pack. In reality, the jet pack was deployed during World War II when the Nazis needed a quick way to transport troops across a river whose bridge had been destroyed. The Nazi jet pack used hydrogen peroxide as fuel, which quickly ignites in contact with a catalyst (such as silver) to release energy and water as waste products. However, there are several problems with jet packs. The main one is that the fuel supply lasts for only thirty seconds to a minute. (In old news clips, you sometimes see daredevils using jet packs to float in the air, such as at the 1984 Olympics. However, these tapes are carefully edited since people float for only thirty seconds to a minute before they fall to the ground.)

The solution to this problem is to develop a portable power pack with enough energy to power longer periods of flight. Unfortunately, no such power supply is available at the present time.

This is also the reason why we don't have ray guns. A laser can work like a ray gun but only if you have a nuclear power plant generating the energy. However, it's impractical to have a nuclear power plant on your shoulders. So jet packs and ray guns will not become real until we create miniature power packs, perhaps in the form of a nanobattery that can store energy at the molecular level.

Another possibility, often featured in paintings and movies of angels or human mutants, is using wings like a bird. On planets with a thick atmosphere, it might be possible to simply jump, flap wings attached to your arms, and take off like a bird. (The thicker the atmosphere, the greater the lift and the easier it is to fly in the air.) So the dream of Icarus could become reality. But birds have several advantages that we don't. Their bones are hollow, and their bodies are quite thin and small compared to their wingspan. Humans, on the other hand, are quite dense and heavy. The human wingspan would have to be twenty to thirty feet across, and we would need much stronger back muscles to flap them. To genetically modify someone to have wings is beyond our technical ability. At present, it is difficult to properly move a single gene, let alone the hundreds of genes necessary to create a viable wing. So while having the wings of an angel is not impossible, the final product is a long way off and will not look like the graceful paintings we are used to seeing.

It was once thought that genetic engineering to modify the human race was the dream of science fiction writers, nothing more. However, a new, revolutionary development has changed all that. The pace of discovery is so dramatic that scientists have hastily convened conferences to discuss slowing down the rate of these new developments.

CRISPR REVOLUTION

The pace of discovery in the field of biotechnology has recently accelerated to a fever pitch with the coming of a new technology called CRISPR (clustered regularly interspaced short palindromic repeats), which promises cheap, efficient, and precise ways to edit DNA. In the past, genetic engineering has been a slow and imprecise process. In gene therapy, for example, a "good gene" is inserted into a virus (which has been neutralized so it is harmless). Then the virus is inserted into the patient, where it quickly infects a person's cells and injects the DNA. The goal is to have the DNA insert itself in the proper place along the chromosome, so that the cell's defective code is replaced by the good gene. Some common diseases are caused by a single misspelling in one's DNA, including sickle-cell anemia, Tay-Sachs, and cystic fibrosis. The hope is this can be corrected.

The results, however, have been disappointing. Often, the body considers the virus to be hostile and mounts a counteroffensive, causing harmful side effects. Also, the good gene often does not implant itself in the correct position. After a fatal incident at the University of Pennsylvania in 1999, many gene therapy experiments were terminated.

CRISPR technology cuts through a lot of these complications. Actually, the basis of the technology evolved billions of years ago. Scientists were puzzled that bacteria developed very precise mechanisms to defeat an onslaught of viruses. How did bacteria recognize a deadly virus and then disarm it? They found that bacteria were able to recognize the threats because they carried a snippet of the virus's genetic material. Like a mug shot, the bacteria were able to use it to identify an invading virus. Once the bacteria recognized the genetic string and therefore the virus, it would cut the virus at a very precise point, neutralizing it and stopping the infection in its tracks.

Scientists were able to replicate this process— successfully replacing a viral sequence with other types of DNA and inserting that DNA in the target cell—making "genomic surgery" possible. CRISPR rapidly replaced older methods of genetic engineering, making gene editing cleaner, more accurate, and much faster.

This revolution took the field of biotechnology by storm. "It just completely changes the landscape," said Jennifer Doudna, one of the pioneers. David Weiss of Emory University said, "All of this has basically happened in a year. It's incredible."

Already, researchers at the Hubrecht Institute in the Netherlands have shown that they can correct a genomic error that causes cystic fibrosis. This raises hopes that many incurable genetic diseases may one day be cured. Many scientists hope that eventually some of the genes for certain forms of cancer can also be replaced using CRISPR technology, thereby stopping the growth of tumors.

Bioethicists, worried about the possible misuses of this technology, have organized conferences to discuss this new science because the side effects and complications are not known, and they made a series of recommendations to try to cool down the furious pace of CRISPR research. In particular, they raised concerns that this technology may lead to germ-line gene therapy. (There are two types of gene therapy, somatic cell gene therapy, where non-sex cells are modified, so that the mutations do not spread to the next generation, and germ-line gene therapy, where your sex cells are altered so that all your descendants can inherit the modified gene.) Germ-line gene therapy could, if unchecked, alter the genetic heritage of the human race. It means that once we venture among the stars, new genetic branches of the human race might emerge. Usually, this would take tens of thousands of years, but bioengineering may reduce that to a single generation if germ-line gene therapy becomes a reality.

In summary, the dreams of science fiction writers who speculated about modifying the human race to colonize distant planets were once considered to be too unrealistic or fanciful. However, with the coming of CRISPR, these far-fetched dreams can no longer be dismissed. Still, we must engage in a thoughtful analysis of all the ethical consequences raised by this fast-moving technology.

ETHICS OF TRANSHUMANISM

These are examples of "transhumanism," which advocates embracing technology to enhance our skills and capabilities. To survive and even flourish on distant worlds, we may have to alter ourselves mechanically and biologically. To transhumanists, it's not a matter of choice but of necessity. Altering ourselves increases our chances on planets with different levels of gravity, atmospheric pressure and composition, temperature, radiation, et cetera.

Rather than being repulsed by technology or fighting its influence, the transhumanists believe that we should embrace it. They relish the idea that we can perfect humanity. To them, the human race was a byproduct of evolution, so our bodies are a consequence of random, haphazard mutations. Why not use technology to systematically improve on these quirks? Their ultimate goal is to create the "posthuman," a new species that can transcend humanity.

Although the concept of altering our genes makes some people squeamish, Greg Stock, a biophysicist affiliated with UCLA, has emphasized that humans have been changing the genetics of the animals and plants around us for thousands of years. When I interviewed him, he pointed out that what appears "natural" to us today is actually a byproduct of intense selective breeding. The modern dinner table would be impossible without the skills of ancient breeders who cultivated plants and animals to suit our needs. (Today's corn, for example, is a genetically modified version of maize and cannot reproduce without human intervention. The kernels, or seeds, do not fall off by themselves, and farmers have to remove and plant them for corn to grow.) And the variety of dogs that we see around us are a byproduct of selectively breeding a single species, the gray wolf. So humans have altered the genes of scores of plants and animals, such as dogs for hunting and cows and chickens for food. In fact, if we could magically remove all the plants and animals that humans have bred over the centuries, our society would look drastically different from what it is today.

As the genes for certain human traits are isolated by scientists, it will be hard to stop people from trying to tinker with them. (For example, if you find out that your neighbor's children have been given genetically enhanced intelligence and they are competing with your children, there will be enormous pressure to have your own kids enhanced in a similar fashion. And in competitive sports, where the rewards are astronomical, it will be exceedingly difficult to stop athletes from trying to enhance themselves.) Whatever ethical hurdles they present, Dr. Stock argues that we shouldn't dismiss genetic enhancements unless a modification is harmful. Or, as Nobel laureate James Watson has said, "No one really has the guts to say it, but if we could make better human beings by knowing how to add genes, why shouldn't we?"

POSTHUMAN FUTURE?

Advocates of transhumanism believe that when we meet advanced civilizations in space, they will have evolved to the point of modifying their biological bodies to accommodate the rigors of living on many different planets. To the transhumanists, advanced civilizations in outer space have most likely achieved a genetically and technologically enhanced future. So if we ever meet aliens from space, we shouldn't be surprised if they are part biological and part cybernetic.

Physicist Paul Davies goes one step further: "My conclusion is a startling one. I think it very likely—in fact inevitable—that biological intelligence is only a transitory phenomenon, a fleeting phase in the evolution of intelligence in the universe. If we ever encounter extraterrestrial intelligence, I believe it is overwhelmingly likely to be post-biological in nature, a conclusion that has obvious and far-reaching ramifications for SETI [the search for extraterrestrial intelligence]."

And AI expert Rodney Brooks wrote, "My prediction is that by the year 2100 we will have very intelligent robots everywhere in our everyday lives. But we will not be apart from them— rather, we will be part robot and connected with the robots."

This debate over transhumanism is actually not a new one but goes back to the last century, when the laws of genetics were first understood. One of the first people to articulate the idea was J. B. S. Haldane, who, in 1923, delivered a lecture, later published in book form, entitled "Daedalus, or Science and the Future," in which he predicted that science could use genetics to improve the condition of the human race.

Many of his ideas seem tame today, but he was aware of the controversy they would generate and admitted that they might appear to be "indecent and unnatural" to someone who read about them for the first time, but that people might eventually accept them.

Finally, the basic principles of transhumanism, that humanity should not have to endure "nasty, brutish, and short" lives, when science can relieve suffering by enhancing the human race, were first clearly laid out by Julian Huxley in 1957.

There are several different views on what aspects of transhumanism we should pursue. Some believe that we should focus on mechanical means to enhance ourselves, such as exoskeletons, special goggles to improve our vision, memory banks that can be uploaded into our brains, and implants to increase our senses. Some believe that genetics should be used to eliminate lethal genes, some believe they should be used to enhance our natural abilities, and some believe they should be used to increase our intellectual powers. Instead of taking decades to perfect certain genetic characteristics via selective breeding, as we have done with dogs and horses, we can accomplish whatever we want in one generation with genetic engineering.

Progress in biotechnology is so rapid that ethical questions abound. And the sordid history of eugenics, including Nazi experiments to create a master race, presents a cautionary tale for anyone interested in altering humans. And it is now possible to take skin cells from a mouse and modify them genetically so that they can become egg and sperm cells, then mate them to produce a healthy mouse. Eventually, this process may be applied to humans. It would vastly increase the number of infertile couples who can successfully produce healthy children, but it also means that people could obtain your skin cells without your permission and create clones of you.

Critics claim that only the rich and powerful will be able to benefit from this technology. Francis Fukuyama of Stanford has warned that transhumanism is "among the world's most dangerous ideas," arguing that if the DNA of our descendants is altered, it will likely change human behavior, create more inequality, and hence undermine democracy. However, the history of technology would indicate that although the wealthy would have early access to these technological miracles, the cost will eventually drop to the point where even the average person can afford them.

Other critics claim that it may be the first step toward splitting the human race and that the very definition of humanity is at stake. Perhaps various branches of genetically enhanced humans will populate different parts of the solar system and eventually diverge into separate species. And one can imagine that rivalries and even warfare may break out between different branches of the human race. Even the concept of "**Homo sapiens**" might be called into question. We will address this important question in chapter 13 when we discuss the world perhaps thousands of years into the future.

In Aldous Huxley's **Brave New World,** biotechnology is used to breed a race of superior beings, called the Alphas, who are destined to lead society from birth. Other embryos are deprived of oxygen, so that they become mentally deficient and hence are bred to serve the Alphas. At the bottom of society are the Epsilons, who are bred to do menial manual labor. This society is a planned utopia that uses technology to satisfy all our needs, and everything appears orderly and peaceful. However, the entire society is based on oppression and misery for those bred to live at the bottom.

Supporters of transhumanism admit that all these hypothetical scenarios must be taken seriously, but at this point they argue that these concerns are purely academic. In spite of the avalanche of new research in biotechnology, much of this talk has to be placed in a larger context. Designer children do not yet exist, and the genes for many personality traits that parents may want for their children have not yet been found. And they may not exist at all. At present, not a single human behavior trait can be changed using biotechnology.

Many argue that the fears of transhumanism run amok are premature, since the technology is still in the distant future. But given the rate at which discoveries are being made, late in this century genetic modifications will probably become a real possibility, so we have to ask the question, How far do we want to take this technology?

CAVEMAN PRINCIPLE

As I've stated in previous books, I believe that the "caveman or cavewoman principle" comes into play and puts a natural limit on how far we will want to alter ourselves. Our basic personality has not changed much since we first emerged as modern humans two hundred thousand years ago. Although today we have nuclear, chemical, and biological weapons, our fundamental desires have remained the same.

And what do we want? Surveys show that after our basic needs are met, we place a high value on the opinions of our peers. We want to look good, especially in front of the opposite sex. We want the admiration of our circle of friends. We might hesitate before altering ourselves too dramatically, especially if it makes us look different from those around us.

Therefore, it is likely that we will only adopt enhancements if they raise our status in society. So though there will be pressure to genetically and electronically enhance our power, especially if we go into outer space and live in different environments, there may well be limits on how much alteration we desire, and that limitation will help keep us grounded.

When Iron Man first appeared in the comics, he was a rather clunky, awkward-looking character. His armor was yellow, round, and ugly. In fact, he looked like a walking tin can. Young children could not identify with him, so the cartoonists later decided to give him a complete makeover. His armor became multicolored, sleek, and form-fitting, clearly enhancing the slim fighting figure of Tony Stark. As a result, his popularity rose dramatically. Even superheroes have to obey the caveman principle.

Golden-era science fiction novels often portray the people of the future as having gigantic bald heads and tiny bodies. Other novels have us evolved into huge brains living in large vats of liquid. But who wants to live like that? I think the caveman principle would prevent us from evolving into creatures that we find repulsive. Instead, we would, in all likelihood, want the ability to increase our life span, memory, and intelligence without having to reform our basic human shape. For example, when playing games in cyberspace, we often have the freedom to choose an animated avatar to represent ourselves. Often we choose avatars that make us attractive or appealing in some way, rather than grotesque or repulsive.

It is also possible that all these technological wonders will backfire and reduce us to helpless children leading pointless lives. In the Disney cartoon **WALL-E,** humans live in spaceships where robots cater to every imaginable whim. Robots do all the heavy lifting and take care of every need, leaving humans with nothing to do but pursue silly pastimes. They become fat, spoiled, and useless and spend their time with idle, pointless pursuits. But I think there is a "baseline" personality that is hardwired into our brains. For example, if drugs are legalized, then many experts estimate that perhaps 5 percent of the human race would become addicted. But the other 95 percent, seeing how drugs can limit or destroy a person's life, will steer clear of them, preferring to live in the real world rather than a drug-altered one. Similarly, once virtual reality is perfected, perhaps a similar number of people may prefer to live in cyberspace rather than in the real world, but it is not likely to be an overwhelming number.

Remember that our cavemen ancestors wanted to be useful and helpful to others. It is hardwired into our genes.

When I first read Asimov's Foundation Trilogy as a child, I was surprised that human beings fifty thousand years from now did not alter themselves. Surely, I thought, humans by then would have completely enhanced bodies, with gigantic heads, tiny withered bodies, and superpowers like in the comic books. But many of the scenes in the novel could have taken place on the Earth today. Looking back at that historic novel, I now realize that the caveman principle was probably at work. I imagine that in the future, people will have the option of putting on devices, implants, and accessories that give them superpowers and enhanced abilities, but afterward they will take most of them off and interact normally in society. Or if they permanently alter themselves, it will be in a way that enhances their standing in society.

WHO DECIDES?

When Louise Brown, the world's first test tube baby, was born in 1978, the technology that made it possible was denounced by many clergymen and columnists, who believed that we were playing God. Today there are more than five million test tube babies in the world; your spouse or best friend may be one.

People have decided to embrace this procedure, in spite of vigorous criticism.

Similarly, when Dolly the Sheep was first cloned in 1996, many critics denounced the technology as being immoral and even profane. Yet today cloning is largely accepted. I asked Robert Lanza, an authority on biotechnology, when the first human cloning might be possible. He pointed out that no one has ever successfully cloned a primate, let alone a human. But he thought that human cloning would one day be possible. And if humans are ever cloned, it is likely only a tiny fraction of the human race will decide to clone themselves. (Perhaps the only people who will clone themselves are rich people who have no heirs, or no heirs they particularly care for. They might clone themselves and give their wealth to themselves as children.)

Some have also denounced "designer children" who are genetically modified by their parents. Yet today, it is commonplace for people to create several in vitro fertilized embryos and then discard the ones that have a potentially fatal mutation (such as Tay-Sachs disease). Thus, in one generation, we might conceivably eliminate these lethal traits from the gene pool.

When the telephone was first introduced in the last century, there were vocal critics. They said it was unnatural to speak to some invisible, disembodied voice in the ether, rather than talking to people face-to-face, and that we would spend too much time on the phone, rather than talking to our children and close friends. The critics, of course, were right. We do spend too much time talking to disembodied voices in the ether. We don't talk to our children enough. But we love the telephone, and sometimes use it to talk to our children. People, rather than editorialists, decided for themselves that they wanted this new technology. In the future, as radical forms of technology that could enhance the human race become available, people will decide for themselves how far to take it. The only way in which these controversial technologies should be introduced is after democratic debate. (Imagine, for the moment, someone from the time of the Inquisition visiting our modern world. Fresh from burning witches and torturing heretics, he might condemn all of modern civilization as being blasphemous.) What seems unethical and even immoral today might seem quite ordinary and mundane in the future.

In any event, if we are to explore the planets and stars, we will have to modify and enhance ourselves to survive the journey. And, since there is a limit to how far we can terraform a distant planet, we will need to adjust ourselves to different atmospheres, temperatures, and gravity. So genetic and mechanical enhancements will be necessary.

But so far, we have only discussed the possibilities of enhancing humanity. What happens when we explore outer space and encounter intelligent life-forms that are completely different from us? Moreover, what happens if we meet civilizations that are millions of years ahead of us in technology?

And if we do not encounter advanced civilizations in outer space, how might we become one ourselves? Although it is impossible to predict the culture, politics, and society of an advanced civilization, there is one thing that even alien civilizations will have to obey, and that is the laws of physics. So what does physics tell us about such advanced civilizations?

Originally, you were clay. From being mineral, you became vegetable. From vegetable, you became animal, and from animal, man . . . And you have to go through a hundred different worlds yet. There are a thousand forms of mind.
—RUMI

If you threaten to extend your violence, this Earth of yours will be reduced to a burnt-out cinder. Your choice is simple: Join us and live in peace or pursue your present course and face obliteration. We shall be waiting for your answer. The decision rests with you.
—KLAATU, ALIEN FROM **THE DAY THE EARTH STOOD STILL**

12 SEARCH FOR EXTRATERRESTRIAL LIFE

One day, the aliens arrived.

They came from distant lands no one had ever heard of, in strange, wondrous ships, using a technology that one could only dream of. They came with armor and shields stronger than anything ever seen before. They spoke an unknown language and brought with them strange beasts.

Everyone was wondering, Who are they? Where do they come from?

Some said they were messengers from the stars.

Others whispered that they were like gods from heaven.

Unfortunately, they were all wrong.

The fateful year was 1519 when Montezuma met Hernán Cortés and the Aztec and the Spanish Empires collided. Cortés and his conquistadors were not messengers from the gods but cutthroats lusting after gold and whatever they could plunder. It took thousands of years for the Aztec civilization to rise from the forest, but, armed with only Bronze Age technology, it was overwhelmed and destroyed by Spanish soldiers in a matter of months.

As we move into outer space, one lesson we can learn from this tragic example is that we should be cautious. The Aztecs, after all, were perhaps only a few centuries behind the Spanish conquistadors in their technology. If we encounter other civilizations in space, they might be so far ahead of us that we can only imagine the power they possess. If we were to enter a war with such an advanced civilization, it might be like King Kong meets Alvin the Chipmunk.

Physicist Stephen Hawking has warned, "We only have to look at ourselves to see how intelligent life might develop into something we wouldn't want to meet." Referring to the consequence of Christopher Columbus meeting Native Americans, he concludes, "That didn't turn out so well." Or, as astrobiologist David Grinspoon says, "If you live in a jungle that might be full of hungry lions, do you jump down from your tree and go, 'Yoo hoo'?"

Hollywood movies, however, have brainwashed us into thinking that we can defeat the alien invaders if they are a few decades or centuries ahead of us in technology. Hollywood assumes that we can win by using some primitive, clever trick. In **Independence Day,** all we have to do is inject a simple computer virus into their operating system to bring them to their knees, as if the aliens use Microsoft Windows.

Even scientists make this mistake, scoffing at the idea that an alien civilization living many light-years away could even visit us. But that assumes that alien civilizations are only a few centuries ahead of us in technology. What happens if they are millions of years ahead of us? A million years is nothing but the blink of an eye in cosmic terms. New laws of physics and new technologies would open up when contemplating these incredible time scales.

Personally, I believe that any advanced civilization in space will be peaceful. They might be aeons ahead of us, which is plenty of time for them to resolve ancient sectarian, tribal, racial, and fundamentalist conflicts. But we must be prepared if they are not. Rather than reaching out and sending radio signals into space to announce our existence to any alien civilization, it might be more prudent to study them first.

I believe we will make contact with an extraterrestrial civilization, perhaps sometime in this century. Instead of being merciless conquerors, they might be benevolent and willing to share their technology with us. This would then be one of the most important turning points in history, comparable to the discovery of fire. It could determine the course of human civilization for centuries into the future.

SETI

Some physicists have actively tried to settle this issue by harnessing modern technology to scan the heavens for signs of advanced civilizations in space. This is called SETI (search for extraterrestrial intelligence) and involves scanning the heavens with the most powerful radio telescopes we possess to listen for transmissions from alien civilizations.

At present, thanks to generous contributions from Paul Allen, cofounder of Microsoft, and others, the SETI Institute is constructing 42 state-of-the-art radio telescopes at Hat Creek, California, about three hundred miles northeast of San Francisco, to scan a million stars. Eventually, the Hat Creek facility may have 350 radio telescopes scanning radio frequencies between one and ten gigahertz.

But working on the SETI project is often a thankless task, begging wealthy donors and wary contributors to fund this project. The U.S. Congress has shown only halfhearted interest, and they finally withdrew all financing in 1993, calling it a waste of taxpayer money. (In 1978, Senator William Proxmire ridiculed it by awarding it his infamous Golden Fleece Award.)

Some scientists, frustrated by the lack of funding, have asked the public to participate directly in order to broaden this search. At the University of California, Berkeley, astronomers created SETI@home, an effort to enlist millions of amateurs online to participate in the search. Anyone can take part. You just download the software from their website. Then at night, while you sleep, your computer searches through the mountain of data they've collected, hoping to find that needle in the haystack.

Dr. Seth Shostak of the SETI Institute in Mountain View, California, whom I have interviewed on a number of occasions, believes that we will make contact with an alien civilization before 2025. I asked him how can he be so sure. After all, decades of hard work have not led to a single verified signal from an alien civilization. Furthermore, using radio telescopes to listen in on alien conversations is a bit of a gamble; maybe the aliens do not use radio. Maybe they use entirely different frequencies, or use laser beams, or an entirely unexpected mode of communication that we haven't thought of. All of these are possible, he admitted. But he was confident that soon we will make contact with alien life. He had the Drake equation on his side.

In 1961, astronomer Frank Drake, not satisfied with all the wild speculation about aliens in space, tried to calculate the odds of finding such a civilization. For example, one can start with the number of stars in the Milky Way galaxy (about one hundred billion) and then reduce that number by the fraction that have planets around them, then by the fraction of the planets that have life on them, then the fraction that have intelligent life, and so on. By multiplying a string of these fractions, one gets a ballpark figure of the possible number of advanced civilizations in the galaxy.

When Frank Drake first proposed this formula, there were so many unknowns that the final results were sheer speculation. Estimates of the number of civilizations in the galaxy ranged from tens of thousands to millions.

Now, however, with the flood of exoplanets found in space, one can make a much more realistic estimate. The good news is that every year, astronomers narrow down the various components of the Drake equation. We now know that at least one out of every five sun-like stars in the Milky Way galaxy has Earth-like planets circling it. According to the equation, we have more than twenty billion such Earth-like planets in our galaxy.

Many more corrections have been made to the

Drake equation. The original equation was too naïve. As we have seen, we now know that Earth-like planets have to be accompanied by Jupiter-sized planets in circular orbits in order to clean out the asteroids and debris that can destroy life. So we have to reduce the number of Earth-like planets to only those that have Jupiter-sized neighbors. Earth-like planets also have to be accompanied by large moons in order to stabilize their spin, or else they will eventually wobble and even flip over after millions of years. (If the moon were tiny, like an asteroid, then small perturbations in the Earth's spin would gradually build up over the aeons, according to Newton's laws, and the Earth might eventually flip over. This would be disastrous for life, since there would be giant earthquakes, monstrous tsunamis, and horrendous volcanic eruptions as the crust of the Earth began to crack. Our moon is large enough so that these perturbations do not build up. But Mars, with tiny moons, may actually have flipped over in the distant past.)

Modern science has given us an ocean of concrete data on how many planets capable of spawning life are out there, but it has also found many more ways in which that life can be extinguished through natural disasters and accidents. There have been many times in the Earth's history when intelligent life was almost extinguished through natural disasters (such as asteroid colli-sions, planet-wide ice ages, volcanic eruptions). A fundamental question is what percentage of planets that meet these criteria actually have life and what percentage of those have escaped plan-etary disasters and spawned intelligent life. So we still are a long way from an accurate assess-ment of the number of intelligent civilizations in our galaxy.

FIRST CONTACT

I asked Dr. Shostak what happens if the aliens come to Earth. Does the president summon an emergency meeting of the Joint Chiefs of Staff? Does the U.N. draft an announcement welcom-ing the aliens? What is the protocol when we make first contact?

His answer was rather surprising: basically, there are no protocols. Scientists have conferences in which they discuss this matter, but they only make informal suggestions that have no official weight. No government takes this issue seriously.

In any event, first contact will likely be a one-way conversation, with a detector on Earth pick-ing up a stray message from a distant planet. But this does not mean that we can establish com-munication with them. Such a signal may come from a star system that is, for example, fifty light-years from Earth, so it will take one hundred years for a message to be sent to that star and a return message sent back to Earth. This means that communication with an ET in space would be extremely difficult.

Assuming that one day they can reach the Earth, a more practical question is, How do we talk to them? What kind of language will they speak?

In the movie **Arrival,** the aliens send huge starships that hover ominously over many nations. When earthlings enter these starships, they are met by aliens who look like gigantic squids. Attempts to interact with them are difficult, since they communicate by scribbling strange characters on a screen, which linguists struggle to translate. A crisis occurs when the aliens scribble a word that can be read either as "tool" or "weapon." Confused by this ambiguity, the nuclear powers put their weapons on high alert. It seems that an interplanetary war is about to break out, all because of a simple linguistic mistake.

(In reality, any species advanced enough to send starships to Earth would probably have been monitoring our TV and radio signals and deciphered our language ahead of time, so they would not have to depend on linguists from Earth. But in any case, it would be unwise to start an interplanetary war with aliens that are perhaps millennia more advanced than we are.)

What happens if the aliens have a totally different frame of reference in their language?

If the aliens descended from a race of intelligent dogs, then their language would reflect smells rather than visual images. If they descended from intelligent birds, their language may be based on complex melodies. If they descended from bats or dolphins, their language may use sonar signals. If they descended from insects, they might signal one another via pheromones.

Indeed, when we analyze the brains of these animals, we see how much they differ from our own brain. While a large portion of our brain is devoted to eyesight and language, the brains of other animals are devoted to things like smell and sound.

In other words, when we make first contact with an alien civilization, we cannot assume that they think and communicate like us.

WHAT DO THEY LOOK LIKE?

When watching a science fiction movie, the highlight is often when we finally see the aliens. (In fact, one of the disappointing features of the otherwise fine movie **Contact** was that, after a tremendous buildup, we never see the aliens themselves.) But in the **Star Trek** series, all the aliens look just like us and talk like us, all speaking perfect American English. The only thing different about them is that they have different types of noses. More imaginative are the aliens in **Star Wars**, who look like wild animals or fish, but they always come from planets where they breathe air and have a gravity similar to Earth's.

At first, one may say that the aliens can look like anything you want, since we have never made contact with them. But there is a certain logic that they are likely to follow. Although we cannot be sure, there is a high probability that life in outer space might begin in the oceans and be composed of carbon-based molecules. Such chemistry is ideally suited to satisfying two vital criteria for life: the ability to store vast amounts of information, because of its complex molecular structure, and the ability to self-replicate. (Carbon has four atomic bonds, which allows it to create long chains of hydrocarbons, which include proteins and DNA. These long carbon DNA chains contain a code in the arrangement of their atoms. These chains occur in two strands, which can unravel and then grab molecules to make a copy of themselves according to this code.)

A new branch of science has recently been born, called exobiology, to study life on distant worlds with ecosystems different from those found on

Earth. So far, exobiologists have had difficulty trying to find a path to creating life-forms that are not based on the carbon chemistry that gives us rich and diverse molecules. Many other possible life-forms have been considered, such as intelligent balloon-like creatures floating in the atmosphere of the gas giants, but it is hard to create a realistic chemistry that makes such creatures possible.

When I was a child, one of my favorite movies was **Forbidden Planet**, which taught me a valuable scientific lesson. On a distant world, astronauts are terrorized by a huge monster that is killing crew members. A scientist takes a plaster mold of the tracks it left behind in the soil. He is shocked by what he finds. The monster's feet, he declares, violate all the laws of evolution. The claws, the toes, bones are all arranged in a way that makes no sense.

This caught my attention. A monster violating the laws of evolution? This was a new concept to me, that even monsters and aliens have to obey the laws of science. Previously, I thought that monsters just had to be ferocious and ugly. But it made perfect sense that monsters and aliens would have to obey the same natural laws that we do. They do not live in a vacuum.

For example, when I hear about the Loch Ness monster, I have to ask what would be the breeding population of such a creature? If a dinosaur-like creature could exist in that lake, it must be part of a breeding population of perhaps fifty or so other creatures. In that case, evidence of these creatures (in the form of bones, carcasses of its prey, waste products, et cetera) should be readily found. The fact that no such evidence has been discovered casts doubt on their existence.

Similarly, one should be able to apply the laws of evolution to aliens in space. It is impossible to tell precisely how an alien civilization might emerge on a distant planet. We can make some inferences, however, based on our own evolution. When we analyze how **Homo sapiens** developed intelligence, we see at least three components that were essential in our rise from the swamp.

1. Some Form of Stereo Eyes

In general, predators are more intelligent than prey. To hunt effectively, one has to be a master of stealth, cunning, strategy, camouflage, and deception. One also has to know the habits of the prey, where they feed, what their weaknesses are, what their defenses are. All this takes some brain power.

On the other hand, all prey have to do is run.

This is reflected in their eyes. Hunters, like tigers and foxes, have eyes facing the front of their face, which gives them stereo vision as the brain compares images from the left and right eyes. This allows them to judge distance, which is essential in locating the prey. However, prey do not need stereo vision. All they need is 360-degree vision to scan for the presence of predators, and hence they have eyes on each side of their face, like deer and rabbits.

In all likelihood, intelligent aliens in space will have descended from predators that hunted for their food. This does not necessarily mean that they will be aggressive, but it does mean that their ancestors long ago might have been predators. We may be well served to be cautious.

2. Some Form of Opposable Thumb or Grasping Appendage

One hallmark of a species that could develop an intelligent civilization is the ability to manipulate the environment. Instead of plants, which are at the mercy of changes in their surroundings, intelligent animals can shape their environment to increase their chances of survival. One thing that set humans apart is the opposable thumb, which gives us the ability to use our hands to exploit tools. Previously, the hand was used mainly to swing from tree branches, and the arc created by our index finger and thumb is roughly the size of a tree branch in Africa. (This does not mean that opposable thumbs are the only grasping instrument that could lead to intelligence. Tentacles and claws may also suffice.)

So the combination of the first and second criteria gives the animal the ability to use hand-eye coordination to hunt for prey and also to manipulate tools. But the third criterion ties it all together.

3. Language

Among most species, any lesson an individual might learn dies with that animal.

In order to hand down and accumulate essential information from generation to generation, some form of language is crucial. The more abstract the language, the more information can be conveyed between generations.

Being a hunter helps encourage the evolution of language, because pack predators have to communicate and coordinate with one another. Language is primarily useful for pack animals. While a single hunter may be crushed by a mastodon, a group of hunters can ambush, surround, trap, snare, and bring down a mastodon. Furthermore, language is necessarily a social phenomenon that accelerates the development of cooperation among individuals. This was an essential ingredient in the rise of human civilization.

I had a graphic example of the social aspects of

language when I swam in a pool full of playful dolphins for a TV program with the Discovery Channel. Inserted in the pool were sonic sensors that recorded the chirps and whistles that they used to communicate with one another. Although they don't have a written language, they have an audible one, which can be recorded and analyzed.

Then, using a computer, one can look for patterns that indicate intelligence. For example, if one randomly analyzes the English language, one finds that the letter **e** is the most common letter in the alphabet. One can then compile a list of letters and analyze how frequently each one is used, providing a distinctive "fingerprint" for that language or particular person. (This can be used to trace the authorship of historical manuscripts—to show, for example, that Shakespeare really did write his plays.)

Similarly, one can record the communications between the dolphins and find that the repetition of their chirps and whistles obeys a mathematical formula.

One can then analyze the language of a number of other species, such as dogs and cats, and find a similar telltale sign of intelligence.

However, as we start to analyze the sounds of insects, we find less and less evidence of intelligence. The point is that animals do have primitive languages, and computers can mathematically calculate their complexity.

EVOLUTION OF INTELLIGENCE ON EARTH

So if at least three attributes are necessary for the development of intelligent life, then we can ask, How many animals on the Earth have all three of them? We find that many predators with stereoscopic vision have claws, paws, fangs, or tentacles but lack the ability to grasp tools. Similarly, none have a sophisticated language that allows them to hunt, share information with others, and hand information down to the next generation.

We can also compare human evolution and intelligence with those of the dinosaurs. Although our understanding of dinosaur intelligence is extremely limited, it is believed that they dominated the Earth for about two hundred million years, yet none of them became intelligent or developed a dinosaur civilization, which took humans only about two hundred thousand years.

But if we analyze the dinosaur kingdom carefully, we see signs that intelligence might have flourished. For example, the velociraptors, immortalized in **Jurassic Park,** probably could have become intelligent with time. They had the stereo eyes of a hunter. They hunted in packs, which meant they probably had some communication system between them to coordinate the hunt. And they had claws for grasping prey, which might have evolved into opposable thumbs. (By contrast, the limbs of the **Tyrannosaurus rex** were tiny, probably used only to grab the flesh after the hunt was over and unlikely to be of much use in grasping tools. The **T. rex** was essentially a walking mouth.)

ALIENS FROM **STAR MAKER**

Given this framework, we can analyze the aliens found in Olaf Stapledon's **Star Maker**. The hero of that story takes an imaginary journey across the universe, encountering scores of fascinating civilizations. We see the panorama of possible intelligence spread across the canvas of the whole Milky Way galaxy.

One alien species evolved on a planet with a large gravitational field. Hence, instead of four legs, they required six legs to walk. Eventually, the front two legs evolved into hands, freeing them up to use tools. Over time, this animal evolved into something resembling a centaur.

He meets aliens that are insect-like. Although each insect is not intelligent, the combination of billions of them creates a collective intelligence. A birdlike race flies in gigantic swarms, like a cloud, and also develops a hive mind. He meets intelligent plantlike creatures that, during the day, are inert like plants but at night can move like animals. He even meets intelligent life-forms that are totally outside our experience, such as intelligent stars.

Many of these alien creatures live in the oceans. One of the most successful of these aquatic species is a symbiosis of two different life-forms, resembling a fish and a crab. With the crab riding behind the fish's head, they can travel rapidly like a fish and the crab can manipulate tools using its claws. This combination gives them a tremendous advantage as they become the dominant species on their planet. Eventually, the crablike creatures venture onto the land, where they invent machines, electrical appliances, rocket ships, and a utopian society based on prosperity, science, and progress.

These symbiotic creatures develop starships and encounter less advanced civilizations. Stapledon writes, "Great care was taken by the Symbiotic race to keep its existence hidden from the primitives, lest they should lose their independence."

In other words, although the fish and the crab separately could not evolve into a higher creature, the combination of the two could.

Given that the vast majority of alien civilizations, if they exist, may live underwater on ice-covered moons (like Europa or Enceladus) or on the moons of rogue planets, the question is, Can an aquatic species become truly intelligent?

If we analyze our own oceans, we see several problems. Fins are an extremely efficient way of traveling in the oceans, while feet (and hands) are not. One can travel and maneuver quite quickly with fins, while moving with feet on the ocean floor is clumsy and awkward. Not surprisingly, in the oceans we see few animals that have evolved appendages that can be used to grasp tools. So creatures with fins are unlikely to become intelligent (unless the fins somehow evolve so they can grasp objects, or else these fins were actually arms and legs of land animals that returned to the oceans, like dolphins and whales).

However, the octopus is a very successful animal. Having survived for at least three hundred million years, it is perhaps the most intelligent of all invertebrates. When we analyze the octopus with regard to our criteria, we find that it matches two out of three of them.

First, being a predator, it has the eyes of a hunter. (However, its two eyes do not focus stereoscopically well toward the front.)

Second, its eight tentacles give it an extraordinary ability to manipulate objects in its environment. These tentacles have remarkable dexterity.

But it has no language to speak of. Being a

solitary hunter, there is no need to communicate with others. There is also no interaction between generations as far as we can tell.

Thus, the octopus displays a certain intelligence. They are notorious for being able to escape from aquariums, taking full advantage of their soft bodies to squeeze through tiny cracks. They also can navigate through mazes, showing they possess some form of memory, and they have been known to manipulate tools. One octopus was able to grab coconut shells and create a shelter for itself.

So if the octopus has some limited intelligence and versatile tentacles, then why didn't it become intelligent? Ironically, it's probably a testament to its success. Hiding under a rock and grabbing prey with its tentacles is a very successful strategy, so octopi probably had no need to develop intelligence. In other words, there was no evolutionary pressure placed on them to evolve greater intelligence.

However, on a distant planet under different conditions, one can imagine that an octopus-like creature could develop a language of chirps and whistles so it could hunt in packs. Perhaps the beak of the octopus can evolve to produce the rudiments of language. One could even imagine that at some point in the distant future evolutionary pressures on Earth could force the octopus to develop intelligence.

So an intelligent race of octopods is certainly a possibility.

Another intelligent creature envisioned by Stapledon was a bird. Scientists have noticed that birds, like the octopods, have significant intelligence. Unlike the octopus, however, they have a very sophisticated way of communicating with one another, through chirps and also through songs and melodies. By recording the songs of certain birds, scientists have noticed that the more sophisticated and melodic the song, the greater the attraction from the opposite sex. In other words, the complexity of a male bird's song allows the female bird to judge its health, its strength, and its suitability as a mate. So there is an evolutionary pressure on them to develop complex melodies and a certain intelligence. Although some birds have the stereo eyes of a hunter (such as hawks and owls) and a form of language, they lack the ability to manipulate the environment.

Millions of years ago, some of the animals that walked on four legs evolved into birds. By analyzing the bones of birds, we see precisely how the bones of the legs slowly evolved into the bones of the wing. There is a one-to-one match between the two sets of bones. But to truly manipulate the environment, one would want animals whose hands are free to grasp tools. This means that intelligent birds will either have to evolve modified wings that have a dual purpose, allowing both flight and the manipulation of tools, or they need to start with at least six legs, four of which eventually become wings and hands.

So an intelligent species of birds is possible, if they could somehow develop the ability to manipulate tools.

These are just a few examples of how varied intelligent species might be. There are certainly many other possibilities one could contemplate.

HUMAN INTELLIGENCE

It is illustrative to ask, Why did we become intelligent? Many primates come close to satisfying all three criteria, so why did we develop these abilities, rather than chimpanzees, bonobos (our closest evolutionary relative), or gorillas?

When we measure **Homo sapiens** against other animals, we see that we are weak and clumsy by comparison. We might easily be the laughing-stock of the animal kingdom. We can't run very fast, we have no talons or claws, we cannot fly, we don't have a keen sense of smell, we have no armor, we aren't very strong, and our skin has no fur and is quite delicate. In every category, we see that there are animals that are vastly superior physically.

In fact, most animals we see around us are very

successful and hence had no evolutionary pressure to change. Some animals haven't changed for millions of years. Precisely because we are weak and clumsy, we were under enormous pressure to acquire skills the other primates lacked. To compensate for our deficiencies, we had to become intelligent.

One theory states that the climate in East Africa began to change several million years ago, causing the forests to recede and the grasslands to spread. Our ancestors were forest creatures, so many of them died off when the trees began to disappear.

Those who did survive were forced to move from the forests onto the savannah and grasslands. They had to wrench their backs and walk upright, allowing them to see above the grass. (We see evidence of this in our swayback, which puts enormous pressure on the small of our back. This is the reason why back problems are one of the most common health issues facing middle-aged people.)

Walking upright had another great advantage: it freed up our hands so we could manipulate tools.

When we encounter intelligent aliens in space, chances are good that they, too, will be clumsy and weak and will have compensated for these deficiencies by evolving intelligence. And they, like us, will evolve the ability to survive by a new technique: the ability to alter their environment at will.

EVOLVING ON DIFFERENT PLANETS

So then how might an intelligent creature develop a modern technological society?

As we've discussed, the most common form of life in the galaxy might be aquatic. We've already looked at whether sea creatures can develop the requisite physiology, but there is a cultural and technological component to our story as well, so let us see if an advanced civilization can rise from the bottom of the ocean.

For humans, after the discovery of agriculture, this process of developing energy and information went through three stages.

The first stage was the industrial revolution, when the energy of our hands was magnified many times by the power of coal and fossil fuels. Society exploded with power, converting a primitive agrarian culture into an industrial one.

The second stage was the electric age, when the power available to us was augmented by electric generators and new forms of communication arose, including radio, TV, and telecommunications. As a result, both energy and information flourished.

The third stage is the information revolution, when computer power came to dominate society.

We can now ask the simple question, Can an aquatic alien civilization also go through these three stages of development in energy and information?

Because Europa and Enceladus are so far from the sun, and since their oceans are perpetually under an ice cover, any intelligent creature on these distant moons will probably be blind, like the fish who live in dark caves beneath the surface of the Earth. Instead, they will probably develop some form of sonar, using sound waves as bats do in order to navigate the oceans.

But since the wavelength of light is so much smaller than the wavelength of sound, it means that they will not be able to see the fine details that we can with our eyes (just as sonograms used by doctors provide far less detail than endoscopy). This will slow down their march toward creating a modern civilization.

But more important, any aquatic species will have a problem with energy, since you cannot burn fossil fuels in water and it's difficult to shield electrical power. Most industrial machinery would be useless without oxygen to create combustion and mechanical motion. Solar power would also be useless, since sunlight would not penetrate the perpetual ice cover.

Without internal combustion engines, fires, and solar power, it would seem that any alien aquatic species would lack the energy to develop into a modern society. There is one source of untapped energy that is available to them, however, and that is geothermal energy coming from heat vents on the bottom of the ocean. Like the volcanic vents on the bottom of our own oceans, similar vents on Europa and Enceladus may provide a convenient energy source for tools.

It might be possible to create an underwater steam engine as well. The temperature of the vents might be well beyond the boiling point of water. If the heat from these vents can be channeled, then creatures might be able to use it to create a steam engine, using a system of pipes that can draw boiling water from these vents and then channel it to move a piston. From this, they might be able to enter the machine age.

It might also be possible to use this heat to melt ores in order to create a metallurgy. If they can extract and mold metal, then they can create cities on the bottom of their oceans. In short, it may be possible to create an underwater industrial revolution.

An electric revolution seems improbable, since water would short-circuit most traditional electrical appliances. Without electricity, all the wonders of that age would be impossible, so their technology would be stunted.

But here also, there is a possible solution. If these creatures can find magnetized iron on the bottom of the oceans, then it's possible to create an electrical generator, which then can be used to power machinery. By spinning these magnets (perhaps by jets of steam hitting a turbine blade), they could push electrons in a wire, creating an electrical current. (This is the same process used in bicycle lamps and hydroelectric dams.) The point is that intelligent underwater creatures might be able to create electric generators using magnets even in the presence of water, and hence enter the electric age.

The information revolution, with computers, is also difficult but not impossible to master for an aquatic species. Just as water is the perfect medium to spawn life, silicon is also the likely basis for any chip-based computer technology. There could be silicon on the bottom of the ocean, which can be mined, purified, and etched to create chips via ultraviolet light, just as we do. (To create silicon chips, UV light is passed through a template that contains the blueprint for all the circuits on a chip. The UV light and a series of chemical reactions creates a pattern that is etched onto a silicon wafer, creating transistors on the chip. This process, which is the basis of transistor technology, can also be done underwater.)

So it would be possible for an aquatic creature to develop intelligence and to create a modern technological society.

NATURAL BARRIERS TO AN ALIEN TECHNOLOGY

Once a civilization begins the long, arduous process of becoming a modern society, it faces yet another problem. There may be a series of natural phenomena that get in the way.

For example, if intelligent creatures evolved on a place like Venus or Titan, they may be faced with a permanent cloud cover over their world, so they would never see the stars. Their concept of the universe would be limited to their planet.

This means that their civilization will never develop astronomy, and their religion would consist of tales that are confined to their planet. Since they will have no urge to explore beyond the clouds, their civilization will also be stunted, and it is highly unlikely they will develop a space program. Without a space program, they would never have telecommunication and weather satellites. (In Stapledon's novel, some creatures living beneath the surface of the sea eventually came onto land, where they discovered astronomy.

If they had stayed in the oceans, they would never have discovered the universe beyond their planet.)

Yet another problem facing a developed society was outlined in Asimov's award-winning story "Nightfall," where he envisioned scientists living on a planet that revolves around six stars. The planet is continually bathed in starlight. Its habitants, who have never seen the night sky with its billions of stars, firmly believe that the entire universe just consists of their solar system. Their entire religion and sense of identity centers around this core belief.

But then scientists begin to make a series of disturbing discoveries. They find that, every two thousand years, their civilization collapses into total chaos. Something mysterious happens to spark the total disintegration of their society. This cycle seems to repeat unending into the past. There are legends that people went insane because everything went dark. People lit huge bonfires to light up the sky, until entire cities went up in flames. Bizarre religious cults spread, governments collapsed, and normal society disintegrated. Then it would take two thousand years before a new civilization could rise from the ashes of the previous one.

Then scientists realize the sickening truth behind their past: that every two thousand years there is an anomaly in their own planet's orbit, so that it experiences nightfall. And to their horror, they find that this cycle will happen again very soon. As the story ends, nightfall begins once again, and civilization descends into chaos.

Stories like "Nightfall" force us to contemplate how life may exist on planets under a totally different set of circumstances from our own. We are lucky to live on the Earth, where energy sources are plentiful, where fire and combustion are possible, where the atmosphere allows electrical devices to function without short circuits, where silicon is plentiful, and where we can see the night sky. If any of these ingredients were missing, it would make the rise of an advanced civilization very difficult.

FERMI PARADOX: WHERE ARE THEY?

But all this still leaves one persistent, nagging question, which is the Fermi paradox: Where are they? If they exist, then surely they would leave a mark, maybe even visit us, yet we see no real evidence of an alien visitation.

There are many possible solutions to this paradox. My thinking is as follows: If they have the ability to actually reach the planet Earth from hundreds of light-years away, then their technology is much more advanced than ours. In that case, we are arrogant to believe that they would travel trillions of miles to visit a backward civilization with nothing to offer. After all, when we visit the forest, do we try to talk to the deer and the squirrels? Maybe initially we might try, but since they don't talk back, we would quickly lose interest and leave.

So for the most part, the aliens would leave us alone, looking at us as a primitive curiosity. Or, as Olaf Stapledon speculated decades ago, perhaps they have a policy not to interfere with primitive civilizations. In other words, they might be aware of us but don't want to influence our development. (Stapledon gives us another possibility as he writes, "Some of these pre-utopian worlds, not malignant but incapable of further advance, were left in peace and preserved, as we preserve wild animals in national parks, for scientific interest.")

When I asked Dr. Shostak this question, he gave me an entirely different answer. He said that a civilization more advanced than ours will most likely develop artificial intelligence, so they would send robots into space. We shouldn't be surprised, he told me, if the aliens that we finally meet are mechanical rather than biological. In movies like **Blade Runner,** robots are sent into outer space to do the dirty work, since space exploration is difficult and dangerous. That, in turn, may explain why we don't pick up their radio emissions. If the aliens follow our own technological path, they will invent robots soon after they invent radio. Once they enter the age of artificial intelligence, they might merge with their robots and have little use for radio anymore.

For example, a civilization of robots may be wired up with cables rather than radio or microwave antennas. Such a civilization would be invisible to the radio receivers of the SETI Project. In other words, an alien civilization may only have a few centuries in which they use radio, so perhaps that is one reason why we don't pick up transmissions.

Others have speculated that maybe they would want to plunder something from our planet. One possibility is the liquid water from our oceans. Liquid water is indeed a precious commodity in our solar system, found only on the Earth and the moons of the gas giants, but ice is not. There is plenty of ice out there on comets, asteroids, and the moons orbiting the gas giants. So all an alien civilization has to do is heat up the ice.

There is another possibility, that maybe they would want to steal valuable minerals from the Earth. This is certainly possible, but there are plenty of uninhabited worlds out there with precious minerals. If an alien civilization has the technology to reach the Earth from vast distances, then they would have a selection of planets to exploit, and it would be far easier to strip a planet that is uninhabited than one with intelligent life.

Another possibility is that they want to steal the heat from the core of the Earth, which would destroy the entire planet. But we suspect that an advanced civilization has harnessed the power of fusion and hence there is no need to steal the heat from the core of the Earth. Hydrogen, the fuel for fusion plants, is after all the most plentiful element in the entire universe. And they can always capture energy from stars, which are also plentiful.

ARE WE IN THEIR WAY?

In **The Hitchhiker's Guide to the Galaxy**, the aliens want to get rid of us because we are simply in the way. The bureaucrats among the aliens have nothing against us personally, but we are an obstacle that had to be removed so they can create an intergalactic bypass. This is a real possibility. For example, who is more dangerous to a deer: a hungry hunter armed with a powerful rifle or a mild-mannered developer with a briefcase who needs land for a housing tract? The hunter may seem more dangerous to a single deer, but ultimately the developer is more lethal to the species, wiping out an entire forest full of creatures.

In the same way, the Martians in **The War of**

the Worlds did not have a grudge against earth-lings. Their world was dying, so they needed to take over ours. They did not hate humans. We were simply in the way.

The same reasoning is found in the previously discussed Superman movie Man of Steel, in which the DNA of the entire population of Krypton was preserved just before their home planet exploded. They need to take over the Earth to resurrect their race. Although this scenario is certainly plausible, again there are other planets to plunder and take over, so one can hope that the aliens would pass us by.

My colleague Paul Davies raises yet another possibility. Maybe their technology is so advanced that they can create virtual reality programs that are far superior to reality, so that they prefer to live perpetually in a fantastic video game. This possibility is not so illogical, because even among humans, a certain fraction of our population would prefer to live in a hazy, drug-fueled state rather than face reality. In our world, this is an unsustainable option, because society would fall apart if everyone were on drugs. But if machines satisfy all our worldly needs, then a parasitic society is possible.

But all this speculation still leaves open the question, What will an advanced civilization, perhaps thousands to millions of years more advanced than ours, look like? Will meeting them usher in a new era of peace and prosperity, or annihilation?

It is impossible to predict the culture, politics, and society of an advanced civilization, but, as I mentioned, there is one thing that even they will have to obey: the laws of physics. So what does physics say about how a super-advanced civilization will evolve?

And if we do not encounter any advanced civilizations in our sector of the galaxy, then how might we advance into the future? Will we be able to explore the stars and eventually the galaxy?

Some scientists have proposed adding the category of a Type IV civilization that controls space-time well enough to affect the entire universe.

Why stop at one universe?
—CHRIS IMPEY

There is something fascinating about science. One gets such wholesale returns of conjecture out of such a trifling investment of fact.
—MARK TWAIN

13 ADVANCED CIVILIZATIONS

The tabloid headlines blared: "Giant Alien Megastructure Found in Space!"

"Astronomers Baffled by Alien Machine in Space!"

Even the **Washington Post**, not used to running lurid stories on UFOs and aliens, ran the headline, "The Weirdest Star in the Sky Is Acting Up Again."

Suddenly, astronomers, who normally analyze boring reams of data from satellites and radio telescopes, were flooded with calls from anxious journalists, asking if it was true that they had finally found an alien structure in space.

This caught them by surprise. The astronomical community was at a loss for words. Yes, something strange had been discovered in space. Yes, it defied explanation, but it was too soon to say what it meant. This might just be a wild goose chase.

The controversy began when astronomers were looking at exoplanets transiting distant stars. Usually, a giant Jupiter-sized exoplanet, moving in front of its mother star, will dim its starlight by 1 percent or so. But one day they were analyzing the data from the Kepler spacecraft concerning the star KIC 8462852, which is about 1,400 light-years from Earth. They found an astonishing anomaly: something had dimmed the starlight by a massive 15 percent in 2011. These anomalies can usually be dismissed. Perhaps there was something wrong with the instruments, a spike in power, a transient surge in electrical output, or perhaps it was nothing but dust on the telescope mirrors.

But then it was observed a second time in 2013, this time dimming the star's light by 22 percent. Nothing known to science can dim starlight regularly by that amount.

"We'd never seen anything like this star. It was really weird," said Tabetha Boyajian, a postdoctoral fellow at Yale.

The situation became even more bizarre when Bradley Schaefer of Louisiana State University searched old photographic plates and found that the star's light has been dimming periodically since 1890. **Astronomy Now** magazine wrote that this "has triggered a frenzy of observations as astronomers hurry to try to get to the bottom of what is rapidly becoming one of the biggest mysteries in astronomy."

So astronomers made long lists of possible explanations. But one by one, doubt was cast on the usual scientific suspects.

What could possibly cause this massive dip in starlight? Could it really be something twenty-two times larger than Jupiter? One possibility was that it was caused by a planet plunging into the star. But that was ruled out because the anomaly kept reappearing. Another possibility was the dust from the disk of the solar system. As a solar system condenses in space, the original disk of gas and dust can be many times larger than the sun itself. So maybe the dimming of starlight occurred because the disk passed in front of the star. But this was ruled out when analyzing the star itself, which was found to be mature. The dust should have long since condensed or been swept into space by the solar winds.

After discarding a number of possible solutions, there was still one option that could not be easily dismissed. No one wanted to believe it, but it could not be ruled out: maybe it was a colossal megastructure built by an alien intelligence.

"Aliens should always be the very last hypothesis you consider, but this looked like something you would expect an alien civilization to build," says Jason Wright, an astronomer from Penn State University.

Since the time elapsed between dips in starlight in 2011 and 2013 was 750 days, astronomers predicted that it would recur again in May 2017. Right on schedule, the star began to dim. This time, practically every telescope on Earth capable of measuring starlight was tracking the star. Astronomers from around the world witnessed the star dimming by 3 percent and then brightening again.

But what could it be? Some thought it might be a Dyson sphere, first proposed by Olaf Stapledon in 1937 but later analyzed by physicist Freeman Dyson. A Dyson sphere is a gigantic sphere around a star, designed to harvest the energy from its massive amounts of starlight. Or it could be a huge sphere orbiting a star that periodically passes in front of the star, causing starlight to dim. Perhaps this was something created in order to power the machines of an advanced Type II civilization. This last supposition tweaked the imagination of amateurs and journalists alike. They asked, What is a Type II civilization?

KARDASHEV SCALE OF CIVILIZATIONS

This classification of advanced civilizations was first proposed by Russian astronomer Nikolai Kardashev in 1964. He was not satisfied looking for alien civilizations without any idea of what he might be searching for. Scientists like to quantify the unknown, so he introduced a scale that ranked civilizations on the basis of energy consumption. Different ones might have different cultures, politics, and history, but all of them would require energy. His ranking was as follows:

1. A Type I civilization utilizes all the energy of the sunlight that falls on that planet.
2. A Type II civilization utilizes all the energy its sun produces.
3. A Type III civilization utilizes the energy of an entire galaxy.

In this way, Kardashev conveniently gave a simple method for computing and ranking the possible civilizations within the galaxy, based on energy use.

Each civilization, in turn, has an energy consumption that can be computed. It is easy to calculate how much sunlight falls on a square foot of land on Earth. Multiplying this by the surface area of the Earth illuminated by the sun and one immediately calculates the approximate energy of an average Type I civilization. (We find that a Type I civilization harnesses the power of 7×10^{17} watts, which is about one hundred thousand times the energy output of the Earth today.)

Since we know the fraction of the sun's energy that falls on the Earth, we can then multiply to include the surface area of the entire sun, and we get its total energy output (which is roughly 4×10^{26} watts). This tells us roughly how much energy is utilized in a Type II civilization.

We also know how many stars there are in the Milky Way galaxy, so we can multiply by this number and find the energy output of an entire galaxy, giving us the energy consumption of a Type III civilization in our galaxy, which is roughly 4×10^{37} watts.

The results were intriguing. Kardeshev found that each civilization was greater than the previous one by a factor of between ten billion and one hundred billion.

One can then mathematically compute when we might rise up this scale. Using the total energy consumption of the planet Earth, we find that we are currently a Type 0.7 civilization.

Assuming a 2 percent to 3 percent annual increase in energy output, which roughly corresponds to the current average growth rate or annual growth in GDP for the planet, we are about a century or two away from becoming a Type I civilization. Rising to the level of a Type II civilization could take a few thousand years, according to this calculation. When we would become a Type III civilization is more difficult to compute, since it involves advances in interstellar travel that are difficult to predict. By one estimate, we will probably not become a Type III civilization for one hundred thousand years and possibly not for a million years.

TRANSITION FROM TYPE 0 TO TYPE I

Of all the transitions, perhaps the most difficult is the transition from Type 0 to Type I, which we are undergoing at present. This is because a Type 0 civilization is the most uncivilized, both technologically and socially. It has risen only recently from the swamp of sectarianism, dictatorship, and religious strife, et cetera. It still has all the scars from its brutal past, which was full of inquisitions, persecutions, pogroms, and wars. Our own history books are full of horrid tales of massacres and genocide, much of it driven by superstition, ignorance, hysteria, and hatred.

But we are witnessing the birth pangs of a new Type I civilization, based on science and prosperity. We see the seeds of this momentous transition germinating every day before our eyes. Already, a planetary language is being born. The internet itself is nothing but a Type I phone system. So the internet is the first Type I technology to develop.

We are also witnessing the emergence of a planetary culture. In sports, we see the rise of soccer and the Olympics. In music, we see the rise of global stars. In fashion, we see the same high-end stores and brands at all the elite malls.

Some fear that this process will threaten local cultures and customs. But in most third-world countries today, the elites are bilingual, fluent in the local language and also a global European language or Mandarin as well. In the future, people will likely be bicultural, fluent in all the customs of the local culture but also at ease with the emerging planetary culture. So the richness and diversity of Earth will survive even as this new planetary culture arises.

Now that we have classified civilizations in space, we can use this to help calculate the number of advanced civilizations in the galaxy. For example, if we apply the Drake equation to a Type I civilization to estimate how plentiful they might be in the galaxy, it would appear they should be quite common. Yet we see no obvious evidence of them. Why? There are several possibilities. Elon Musk has speculated that, as civilizations master advanced technology, they develop the power to destroy themselves and that the biggest threat facing a Type I civilization may be a self-inflicted one.

For us, there are several challenges as we make the transition from Type 0 to Type I: global warming, bioterrorism, and nuclear proliferation, to name a few.

The first and most immediate is nuclear proliferation. The bomb is spreading into some of the most unstable regions of the world, such as the Middle East, the Indian subcontinent, and the Korean peninsula. Even small countries may one day have the ability to develop nuclear weapons. In the past, it took a large nation-state to refine uranium ore into weapons-grade materials. Gigantic gaseous diffusion plants and banks of ultracentrifuges were required. These enrichment facilities were so large they could easily be seen by satellite. This was beyond the reach of small nations.

But blueprints for nuclear weapons have been stolen and then sold to unstable regimes. The cost of ultracentrifuges and purifying uranium into weapons-grade material has fallen. As a result, even nations like North Korea, which is perpetually teetering on the brink of collapse, can amass a small but deadly nuclear arsenal today.

Now the danger is that a regional war, between India and Pakistan, say, could escalate to a major war, drawing in the major nuclear powers. Since the United States and Russia each possess about seven thousand nuclear weapons, this threat is significant. There is even a concern that nonstate actors or terrorist groups could procure a nuclear bomb.

The Pentagon commissioned a report from the

Global Business Network think tank that analyzed what might happen if global warming destroys the economies of many poor nations such as Bangladesh. It concluded that, in a worst-case scenario, nations may use nuclear weapons to protect their borders from being overrun by a flood of millions of desperate, starving refugees. And even if it does not cause a nuclear war, global warming is an existential threat to humanity.

GLOBAL WARMING AND BIOTERRORISM

Since the end of the last glacial period about ten thousand years ago, the Earth has been gradually warming up. However, over the past half century, the Earth has been heating at an alarming and accelerating rate. We see evidence of this on numerous fronts:

- Every major glacier on the Earth is receding
- The northern polar ice has thinned by an average of 50 percent over the past fifty years
- Large parts of Greenland, which is covered by the world's second-largest ice sheet, are thawing out
- A section of Antarctica the size of Delaware, the Larsen Ice Shelf C, broke off in 2017,

and the stability of the ice sheets and ice shelves is now in question

• The last few years have been the hottest ever recorded in human history
• The Earth's average temperature has increased by about 1.3 degrees Celsius in the past century
• On average, summer is about one week longer than it was in the past
• We are seeing more and more "one-hundred-year events," such as forest fires, floods, droughts, and hurricanes

There is the danger that, if this global warming accelerates unabated into the coming decades, it could destabilize the nations of the world, create mass starvation, generate mass migration from the coastal areas, and threaten the world economy and prevent the transition to a Type I civilization.

There is also the threat of weaponized bio-germs that could potentially wipe out 98 percent of the human population.

Throughout world history, the greatest killers have not been wars but plagues and epidemics. Unfortunately, it is possible that nations have kept secret stockpiles of deadly diseases, such as small-pox, which could be weaponized using biotech-nology to create havoc. There is also the danger that someone could create a doomsday weapon by bioengineering some existing disease—Ebola, HIV, avian flu—and making it more lethal or causing it to spread more quickly and easily.

Perhaps in the future, if we ever venture to other planets, we may find the ashes of dead civilizations: planets whose atmospheres are highly radioactive; planets that are too hot, because of a runaway greenhouse effect; or planets with empty cities because they used advanced biotech weaponry on themselves. So the transition from Type 0 to Type I is not guaranteed and in fact represents the greatest challenge facing an emerging civilization.

ENERGY FOR TYPE I CIVILIZATION

A key question is whether a Type I civilization can make the transition to energy sources other than fossil fuels.

One possibility is to harness uranium nuclear power. But uranium fuel for a conventional nuclear reactor creates large amounts of nuclear waste products, which are radioactive for millions of years. Even today, fifty years into the nuclear age, we still do not have a safe way to store high-level nuclear waste. This material is also quite hot and can create a meltdown, as we have seen in the Chernobyl and Fukushima disasters.

An alternative to uranium fission power is fusion power, which, as we saw in chapter 8, is not ready yet for commercial use, but a Type I civilization a century more advanced than ours may have perfected the technology and could use it as an indispensable source of nearly unlimited energy.

One advantage of fusion power is that its fuel is hydrogen, which can be extracted from seawater. A fusion plant also cannot suffer a catastrophic meltdown like the ones we saw at Chernobyl and Fukushima. If there is a malfunction in the fusion plant (such as the superhot gas touching the lining of the reactor) the fusion process automatically shuts itself off. (This is because the fusion process has to attain the Lawson criterion: it must maintain the proper density and temperature to fuse the hydrogen over a certain period of time. But if the fusion process gets out of control, the Lawson criterion is no longer satisfied, and it stops by itself.)

Also, a fusion reactor only produces modest amounts of nuclear waste. Because neutrons are created in the process of fusing hydrogen, these neutrons can irradiate the steel of the reactor, making it slightly radioactive. But the amount of waste created in this fashion is only a tiny fraction of that generated by uranium reactors.

In addition to fusion power, there are other

possible renewable energy sources. One attractive possibility for a Type I civilization is to exploit space-based solar energy. Since 60 percent of the energy of the sun is lost passing through the atmosphere, satellites could harness much more solar energy than collectors on the surface of the Earth.

A space-based solar energy system might consist of many huge mirrors orbiting the Earth collecting sunlight. They would be geostationary (orbiting the Earth at the same rate at which the Earth rotates, so they appear to be in a fixed location in the sky). This energy can then be beamed down to a receiving station on the Earth in the form of microwave radiation, and it would then be distributed through a traditional electrical grid.

There are many advantages to space solar energy. It is clean and without waste products. It can generate power twenty-four hours a day, rather than just during daylight hours. (These satellites are almost never in the shadow of the Earth, since their path takes them considerably away from the Earth's orbit.) The solar panels have no moving parts, which vastly reduces breakdowns and repair costs. And best of all, space solar power taps into a limitless supply of free energy from the sun.

Every scientific panel that has looked into the

question of space solar has concluded that the goal is achievable with off-the-shelf technology. But the main problem, like all endeavors involving space travel, is cost. Simple estimates show that this is currently many times more expensive than simply putting solar panels out in your backyard.

Space solar energy is beyond the means of a Type 0 civilization like ours, but it may become a natural source of energy for a Type I civilization for several reasons:

1. The cost of space travel is dropping, especially because of the introduction of private rocket companies and the invention of reusable rockets.
2. The space elevator may be possible late in this century.
3. Space solar panels can be made of lightweight nanomaterials, keeping weight and costs down.
4. The solar satellites can be assembled in space by robots, eliminating the need for astronauts.

It also is generally considered safe because while microwaves can be harmful, calculations show that most of the energy is confined within the beam, and the energy that escapes outside the beam should fall within accepted environmental standards.

TRANSITION TO TYPE II

Eventually, a Type I civilization may exhaust the power available on its home planet and look to exploit the enormous energy found in the sun itself.

A Type II civilization should be easy to find, because they are likely immortal. Nothing known to science can destroy their culture. Meteor or asteroid collisions can be avoided using rocketry. The greenhouse effect can be avoided using hydrogen-based or solar technologies (fuel cells, fusion plants, space solar satellites, et cetera). If there are any planetary threats, they can even leave their home in large space armadas. They might even be able to move their planet if necessary. Since they have enough energy to deflect asteroids, they can whip them around their planet, causing a small shift in its trajectory. With successive "slingshot" maneuvers, they could move the orbit of their planet farther from the sun if their star is late in its life cycle and beginning to expand.

To supply energy for their civilization, they might, as we mentioned earlier, build a Dyson sphere to harvest most of the energy from the sun itself. (One problem with building such gigantic megastructures is there might not be enough building material on the rocky planets to construct them. Since our sun is 109 times bigger than the Earth in diameter, it would require an immense amount of material to build one of these structures. Perhaps the solution to this practical problem is to use nanotechnology. If these megastructures are made of nanomaterials, they might only be a few molecules in thickness, which would vastly decrease the amount of building materials required.)

The number of space missions needed to create such megastructures is truly monumental. But the key to building them may be to utilize space-based robots and self-organizing materials. For example, if a nanofactory could be built on the moon to make panels for the Dyson sphere, they could be assembled in outer space. Because these robots are self-replicating, an almost unlimited number of them could be built to create this structure.

But even if a Type II civilization is virtually immortal, it still faces a long-term threat: the second law of thermodynamics, the fact that all their machines will create enough infrared heat radiation to make life impossible on their planet. The second law says that entropy (disorder, chaos, or waste) always increases in a closed system. In this case every machine, every appliance, every apparatus generates waste, in the form of heat. Naïvely, we can assume that the solution is to build gigantic refrigerators to cool down the planet. These refrigerators do in fact lower the temperature inside them, but if we add everything up, including heat from the motors used by the refrigerators, the average heat of the whole system still increases.

(For example, on a very hot day, we fan our faces for relief, thinking that this cools us down. Fanning ourselves does cool down our face, giving us temporary relief, but the heat generated by the motion of our muscles, bones, and so on actually produces more net heat. So fanning ourselves gives us immediate psychological relief, but our total body temperature and the temperature of the air around us actually go up.)

COOLING DOWN A TYPE II CIVILIZATION

A Type II civilization, in order to survive the second law, may necessarily have to disperse its machinery or overheat. As we discussed earlier, one solution would be to move most of the machinery to outer space, so that the mother planet becomes a park. This means that a Type II civilization might build all its heat-generating equipment off the planet. Although it consumes the energy output of a star, the waste heat generated is in outer space and hence dissipates harmlessly.

Eventually the Dyson sphere itself begins to heat up. This means that a Dyson sphere must necessarily emit infrared radiation. (Even if we assume that the civilization creates machines to try to conceal this infrared radiation, eventually these machines themselves become hot and radiate in the infrared.)

Scientists have scanned the heavens looking for the telltale signs of infrared radiation from a Type II civilization, and they have failed to find it. Scientists at Fermilab outside Chicago scanned 250,000 stars looking for signatures of a Type II civilization but only found four that were "amusing but still questionable," so their results were inconclusive. It is possible that the James Webb Space Telescope, which will go into service late in 2018 and will look specifically for infrared radiation, may have the sensitivity to find the heat signature of all Type II civilizations in our sector of the galaxy.

So this is a mystery. If Type II civilizations are virtually immortal, and they necessarily emit waste infrared radiation, then why haven't we detected them? Perhaps looking for infrared emissions is too narrow.

Astronomer Chris Impey of the University of

Arizona, commenting on finding a Type II civilization, has written, "The premise is that any highly advanced civilization will leave a much larger footprint than we will. Type II or later civilizations may employ technologies that we're tinkering with or can barely imagine. They might orchestrate stellar cataclysms or use propulsion by anti-matter. They might manipulate spacetime to create wormholes or baby universes and communicate by gravity waves."

Or, as David Grinspoon has written, "Logic tells me that it is reasonable to look for godlike signs of advanced aliens in the sky. And yet the idea seems ridiculous. It is both logical and absurd. Go figure."

One possible way out of this dilemma is to realize that there are two ways to rank a civilization: by its energy consumption, but also by its information consumption.

Modern society has expanded in the direction of miniaturization and energy efficiency as it consumes an exploding amount of information. In fact, Carl Sagan proposed a way to rank civilizations by information.

In this scenario a Type A civilization consumes a million bits of information. A Type B civilization would consume ten times that number, or ten million bits of information, and so on, until we hit Type Z, which can consume an astounding 10^{31} bits of information. By this calculation, we are a Type H civilization. The point here is that civilizations may advance on the scale of information consumption while consuming the same amount of energy. Thus they may not produce a significant amount of infrared radiation.

We see an example of this when we visit a science museum. We are amazed at the size of the machines of the industrial revolution, with gigantic locomotives and huge steamboats. But we also notice how inefficient they were, generating a large amount of waste heat. Similarly, the gigantic computer banks of the 1950s can be surpassed by an ordinary cell phone today. Modern technology became much more sophisticated, intelligent, and less wasteful of energy.

So a Type II civilization can consume a vast amount of energy without burning up by distributing their machines in Dyson spheres, on asteroids and nearby planets, or by creating superefficient miniaturized computer systems. Instead of being consumed by the heat generated by their huge energy usage, their technology may also be superefficient, consuming vast amounts of information and producing relatively little waste heat.

WILL HUMANITY SPLIT APART?

There are limitations, however, to how far each civilization will advance in terms of space travel. For example, a Type I civilization, as we have seen, is limited by its planetary energy. At best, it will master the art of terraforming a planet like Mars and begin to explore the nearest stars. Robotic probes will begin exploring nearby solar systems and perhaps the first astronauts will be sent to the nearest star, like Proxima Centauri. But its technology and its economy are not sufficiently advanced to begin the systematic colonization of scores of nearby star systems.

For a Type II civilization, which is centuries to millennia more advanced, colonization of a sector of the Milky Way becomes a real possibility. But even for a Type II civilization, eventually they are constrained by the light barrier. If we assume that faster-than-light propulsion is not available to them, it may take many centuries to colonize their sector of the galaxy.

But if it takes centuries to go from one star system to another, then eventually the ties to the home world become extremely tenuous. Planets will eventually lose contact with other worlds, and new branches of humanity may emerge that can adapt to radically different environments. Colonists may also genetically and cybernetically modify themselves to adapt to strange environments. Eventually, they may not feel any connection to the home planet.

This seems to contradict the vision of Asimov in his Foundation series, with a Galactic Empire emerging fifty thousand years from now that has colonized most of the galaxy. Can we reconcile these two very different visions of the future?

Is the ultimate fate of human civilization to splinter into smaller entities, with only the sketchiest knowledge of one another? This raises the ultimate question: Will we gain the stars but lose our humanity in the process? And what does it mean to be human anyway if there are so many distinct branches of humanity?

This divergence seems to be universal in nature, a common thread that runs through all of evolution, not just humanity. Darwin was the first to see how this occurs through the animal and plant kingdoms when he sketched a prophetic diagram in his notebook. He drew a picture of the branches of a tree, with different arms diverging into smaller branches. In one simple diagram, he drew the tree of life, with all the diversity of nature evolving from a single species.

Perhaps this diagram applies not only to life on Earth but to humanity itself thousands of years from now, when we become a Type II civilization capable of colonizing the nearby stars.

GREAT DIASPORA IN THE GALAXY

To gain some concrete insight into this problem, we have to reanalyze our own evolution. Looking at the sweep of human history, we can see that roughly seventy-five thousand years ago, a Great Diaspora took place, with small bands of humans moving away from Africa through the Middle East, creating settlements along the way. Perhaps driven by ecological disasters, such as the Toba eruption and a glaciation period, one of the main branches went through the Middle East and journeyed on to Central Asia. Then this migration split further into several smaller branches about forty thousand years ago. One branch kept on going east and eventually settled in Asia, forming the core of the modern Asian people. The other branch turned around and went into northern Europe, eventually becoming Caucasians. Yet another branch went southeast and eventually passed through India and into Southeast Asia and then Australia.

Today, we see the consequences of this Great Diaspora.

We see a variety of humans of different colors, sizes, shapes, and cultures who have no ancestral memory of their true origins. One can even calculate roughly how divergent the human race is. If we assume that one generation is 20 years long, then at most about 3,500 generations separate any two humans on the planet.

But today, tens of thousands of years later, with modern technology, we can begin to re-create all the migration routes of the past and build an ancestral family tree of human migrations over the past seventy-five thousand years.

I had a vivid demonstration of this while hosting a BBC TV science special about the nature of time. BBC took some of my DNA and sequenced it. Four of my genes were then carefully compared with the genes of thousands of other individuals around the world, looking for a match. Then the locations of the people who matched these four genes were identified on a map. The result was rather interesting. It showed a concentration of people scattered through Japan and China who had a match, but then there was a thin trail of dots that tapered off into the distance near the Gobi Desert, through Tibet. So, using DNA analysis, it was possible to retrace the route that my ancestors took about twenty thousand years ago.

HOW FAR WILL WE DIVERGE?

How far will humanity diverge over thousands of years? Will humanity be recognizable after tens of thousands of years of genetic separation?

This question can actually be answered using DNA as a "clock." Biologists have noticed that DNA mutates at roughly the same rate across the ages. For example, our closest evolutionary neighbor is the chimpanzee. Analysis of the chimpanzee shows that we differ by approximately 4 percent of our DNA. Studies of chimpanzee and human fossils indicate that we separated from them about 6 million years ago.

This means that our DNA mutated at the rate of 1 percent over a period of 1.5 million years. This is only an approximate number, but let us see if it can allow us to understand the ancient history of our own DNA.

Assume, for the moment, that this rate of change (1 percent change every 1.5 million years) is roughly constant.

Now let us analyze the Neanderthal, our closest humanlike kin. DNA and fossil analysis of the Neanderthal show that their DNA differs from our DNA by about 0.5 percent and that we separated from them roughly five hundred thousand to a million years ago. So this is in rough agreement with the DNA clock.

If we now analyze the human race, we find that any two humans chosen at random can differ in their DNA by 0.1 percent. Our clock then says that different branches began to diverge about 150,000 years ago, which is in rough agreement with the actual origins of humanity.

So given this DNA clock, we can calculate roughly when we diverged from the chimpanzees, the Neanderthals, and also our fellow human beings.

The point is that we can use this clock to estimate how far humanity will change in the future if we disperse throughout the galaxy and don't drastically tinker with our DNA. Assume for the moment that we remain a Type II civilization with only sub-light-speed rockets for 100,000 years.

Even if different human settlements lose all contact with other branches of humanity, this means humans will probably only diverge by about 0.1 percent in our DNA, which is the amount of divergence that we already see today among humans.

The conclusion here is that, as humanity spreads throughout the galaxy at sub-light speed and different branches lose all contact with other branches, we will still be basically human. Even after 100,000 years, when we might reasonably be expected to attain light speed, different human settlements will differ no more than any two humans on the Earth today.

This phenomenon also applies to the very language that we speak. Archeologists and linguists have noticed that a startling pattern emerges when they try to trace the origin of language. They find that languages constantly branch out into other smaller dialects due to migrations; over time, these new dialects become full-fledged languages themselves.

If we create a vast tree of all known languages and how they branched off one another and compare it with the ancestral tree detailing ancient migration routes, we find an identical pattern.

For example, Iceland, which has been largely isolated from Europe since 874 AD, when the first Norwegian settlements began, can be used as a laboratory to test linguistic and genetic theories. The Icelandic language is closely related to the Norwegian language of the ninth century, with a little bit of Scottish and Irish thrown in the mix. (This is probably due to the Vikings taking slaves from Scotland and Ireland.) It is then possible to create a DNA clock and a linguistic clock to roughly calculate how much divergence there is over a thousand years. Even after a thousand years, one can easily find evidence of ancient migration patterns imprinted in their language.

But even if our DNA and language still resemble themselves after thousands of years of separation, what about our culture and our beliefs? Will we be able to understand and identify with these divergent cultures?

COMMON CORE VALUES

When we look at the Great Diaspora and the civilizations that it created, we see not only a variety of physical differences in skin color, size, hair, et cetera, but also a certain core set of characteristics that are remarkably the same across all cultures, even when they lost all contact with one another for thousands of years.

We see evidence of this today when we go to the movies. People of different races and cultures, who might have diverged from us seventy-five thousand years ago, still laugh, cry, and thrill at the same moment in the film. Translators of foreign films notice the commonality of the jokes and humor in the movies, although the languages themselves diverged long ago.

This also applies to our sense of aesthetics. If we visit an art museum that has exhibits from ancient civilizations, we see common themes. Regardless of the culture, we find artwork depicting landscape scenes, portraits of the rich and powerful, and images of myths and gods. Although the sense of beauty is difficult to quantify, what is considered beautiful in one culture is often considered beautiful by another totally unrelated culture. For example, no matter which culture we examine, we see similar flowers and floral patterns.

Another theme that cuts across the barriers of space and time is our common social values. One core concern is for the welfare of others. This means kindness, generosity, friendship, thoughtfulness. Various forms of the Golden Rule are found in numerous civilizations. Many of the religions of the world, at the most fundamental level, stress the same concepts, such as charity and sympathy for the poor and unfortunate.

Thus, the caveman principle recognizes that our core personalities have not changed much in two hundred thousand years, so even as we spread out among the stars, we will most likely retain our values and personal characteristics.

Furthermore, psychologists have noted that there might be an image of what is attractive that is encoded in our brain. If we take photographs of hundreds of different people at random and then, using computers, superimpose these pictures on top of one another, we see a composite, average image that emerges. Surprisingly, this image is considered by many to be attractive. If true, this implies that there is an average image that might be hardwired in our brains that determines what we consider to be attractive. What we consider to be beautiful in a person's face is actually the norm, not the exception.

But what happens when we finally attain Type III status and have the capability of faster-than-light travel? Will we spread the values and aesthetics of our world across the galaxy?

TRANSITION TO TYPE III

Eventually, a Type II civilization may exhaust the power of not just its home star but all the nearby stars and gradually start the journey to become a Type III civilization, which is galactic. Not only can a Type III civilization harvest the energy from billions of stars, it can also harness the energy of black holes, like the supermassive one located at the center of the Milky Way galaxy, which weighs as much as two million suns. If a starship travels in the direction of our galactic nuclei, we find a vast collection of dense stars and dust clouds that would be an ideal source of energy for a Type III civilization. To communicate across the galaxy, such an advanced civilization may use gravity waves, which were first predicted by Einstein in 1916 but finally detected by physicists in 2016. Unlike laser beams, which might be absorbed, scattered, and diffused as they travel, gravity waves would be able to spread across the stars and galaxy and therefore may be more reliable over great distances.

It is unclear at this point whether faster-than-light travel is feasible, so we need to consider for the moment the possibility that it is not.

If only sub-light spacecraft are possible, then a Type III civilization may decide to explore the billions of worlds in their galactic backyard by sending self-replicating probes that travel at sub-light speeds to the stars. The idea is to place these robotics on a distant moon. Moons make an ideal choice because their environments are more stable, without erosion, and they are easy to land on and leave from, because of their low gravity. With solar collectors to supply energy, a lunar probe can scan the solar system and radio back useful information indefinitely.

Once it has landed, the probe will create a factory from the lunar material in order to manufacture a thousand copies of itself. Each clone in the second generation then blasts off to colonize other distant moons. So, starting with one robot, we then have a thousand. If each of them creates another thousand robots, then we have a million. Then a billion. Then a trillion. In just a few generations, we can have an expanding sphere containing quadrillions of these devices, which scientists call von Neumann machines.

This in fact is the plot of the movie **2001**, which even today portrays perhaps the most realistic encounter with an alien intelligence. In that movie, aliens put a von Neumann machine, the monolith, on the moon, which sends signals to a relay station based on Jupiter in order to monitor and even influence the evolution of humanity.

So our first encounter may not be with a bug-eyed monster but with a small self-replicating probe. This could be quite small, miniaturized by nanotechnology, perhaps so small that you would not even notice it. Conceivably, in your backyard or on the moon, there is evidence of a past visitation that is nearly invisible.

In fact, Professor Paul Davies has made a proposal. He wrote an article advocating going back to the moon in order to search for anomalous energy signatures or radio transmissions. If a von Neumann probe landed on the moon millions of years ago, it would likely use sunlight for its power, so it could continually broadcast radio emissions. And since the moon has no erosion, chances are it will be in near-perfect working condition and may still be in operation.

Since there is renewed interest in going back to the moon and then on to Mars, this would give scientists an excellent opportunity to see if any evidence exists for the presence of previous visitations.

(Some people, like Erich von Däniken, have claimed that alien ships already landed centuries ago and that these alien astronauts are depicted in the artwork of ancient civilizations. They claim that the elaborate headdresses and costumes often found in ancient paintings and monuments are actually depictions of ancient astronauts, with their helmets, fuel tanks, pressure suits, et cetera. While this idea cannot be dismissed, it is very difficult to prove. Ancient paintings are not enough. We need positive, tangible proof of previous visitations. For example, if there were alien spaceports, there must be debris and waste left over, in the form of wires, chips, tools, electronics, garbage, and machinery. One alien chip would settle this entire debate. So if one of your acquaintances claims to have been abducted by aliens from space, tell him or her to steal something from the ship the next time it happens.)

So even if light speed cannot be broken, a Type III civilization could have trillions upon trillions of probes spread across the entire galaxy within a few hundred thousand years, all sending back useful information.

Von Neumann machines may be the most efficient way for a Type III civilization to obtain information concerning the state of the galaxy. But there is yet another way to explore the galaxy more directly, and this is through something I call "laser porting."

LASER PORTING TO THE STARS

One of the dreams of science fiction writers is to be able to explore the universe as pure-energy beings. Perhaps one day, far in the future, we might be able to shed our material existence and roam the cosmos, riding on a beam of light. We would be able to travel to distant stars at the fastest possible velocity. When we are free of material constraints, we would be able to ride alongside comets, skim the surface of erupting volcanoes, fly past the rings of Saturn, and visit destinations on the other side of the galaxy.

Instead of being a flight of fantasy, this dream may actually be rooted in solid science. In chapter 10, we analyzed the Human Connectome Project, the ambitious effort to map the entire brain. Perhaps late in this century or early in the next, we will have the complete map, which in principle will contain all our memories, sensations, feelings, even our personality. Then the connectome might be placed on a laser beam and sent into outer space. All the information necessary to create a digital copy of your mind can travel across the heavens.

In one second, your connectome could be sent to the moon. Within minutes it could reach Mars. Within hours, it could reach the gas giants. And within four years, you could visit Proxima

Centauri. Within a hundred thousand years, you could reach the ends of the Milky Way galaxy.

Once it arrives on a distant planet, the information on the laser beam would be downloaded into a mainframe computer. Then your connectome could control a robotic avatar. Its body is so sturdy that it can survive even if the atmosphere is poisonous, the temperature is freezing or hellish, or the gravity strong or weak. So although all your neural patterns are contained inside the mainframe computer, you have all the sensations coming from the avatar. For all intents and purposes, you are inhabiting it.

The advantage of this approach is that there is no need for messy, expensive booster rockets or space stations. You never face the problem of weightlessness, asteroid collisions, radiation, accidents, and boredom because you are transmitted as pure information. And at the speed of light, you have taken the fastest possible journey to the stars. From your point of view, the trip is instantaneous. All you remember is entering the laboratory and then instantly arriving at your destination. (This is because time effectively stops while riding on the light beam. Your consciousness is frozen as you move at the speed of light, so you travel across the cosmos without any time delay. This is quite different from suspended animation, since when traveling at the speed of light, as I mentioned, time effectively stops. And while you would not see the sights while you were in transit, you could stop at any relay station and observe your surroundings.)

I call this "laser porting," and it is perhaps the most convenient and rapid way to reach the stars. A Type I civilization a century from now may be able to conduct the first laser porting experiments. But for Type II and III civilizations, laser porting may be the preferred method of transportation across the galaxy because they will most likely have already colonized distant planets with self-replicating robots. Perhaps a Type III civilization would have a vast laser porting superhighway connecting the stars in the Milky Way galaxy with trillions of souls in transit at any one time.

Although this idea seems to provide the most convenient way to explore the galaxy, to actually create the laser port requires solving several practical problems.

Placing your connectome on a laser beam is not a problem, since lasers can in principle transport unlimited amounts of information. The main problem is to create a network of relay stations along the way that receive your connectome, amplify it, and send it along to the next station. As we mentioned, the Oort Cloud extends several light-years from a star, so the Oort Clouds from different stars can overlap. Thus, stationary comets in the Oort Cloud may provide ideal sites for these relay stations. (Creating relay stations on Oort Cloud comets would be preferable to placing them on a distant moon, since moons orbit around planets and are often obscured by them, while these comets are stationary.)

As we've seen, these relay stations can only be set up at slower-than-light-speed velocities. One way to solve this problem is to use a system of laser sails, which travel at a significant fraction of the speed of light. Once these laser sails land on an Oort Cloud comet, they could use nanotechnology to make copies of themselves and assemble a relay station using the raw materials found on the comet.

So although the original relay stations would have to be made at sub-light speeds, after that our connectomes could be free to roam at light speed.

Laser porting could be used not only for scientific purposes but also for recreation. We might take a vacation among the stars. We would first map out a sequence of planets, moons, or comets we wish to visit, no matter how hostile or dangerous the environment may be. We might make a checklist of the types of avatars that we wish to inhabit. (These avatars do not exist in virtual reality but are actual robots endowed with superhuman powers.) So on each planet, there is an avatar waiting for us with all the traits and super-powers we desire. When we reach that planet, we assume the identity of that avatar, travel across the planet, and enjoy all the incredible sights. Afterward, we return the robot for the next customer to use. Then we laser-port to the next destination. In a single vacation, we may be able to explore several moons, comets, and exoplanets. We never have to worry about accidents or illnesses, since it is just our connectome roaming across the galaxy.

So when we gaze into the heavens at night wondering if anyone is out there, although it may appear to be cold, still, and empty, perhaps the night sky is teeming with trillions of travelers being sent at the speed of light across the heavens.

WORMHOLES AND THE PLANCK ENERGY

This, however, leaves open the second possibility, that faster-than-light travel might be possible for a Type III civilization. A new law of physics enters into this picture. This is the realm of the Planck energy, the scale at which bizarre new phenomena occur that violate the usual laws of gravity.

To understand why the Planck energy is so important, it is essential to realize that at present all known physical phenomena, from the Big Bang to the motion of subatomic particles, can be explained by two theories: Einstein's general theory of relativity and the quantum theory. Together, they represent the bedrock physical laws governing all matter and energy. The first, general relativity, is the theory of the very big: relativity explains the Big Bang, the properties of black holes, and the evolution of the expanding universe. The second is the theory of the very small: the quantum theory describes the properties and motion of atomic and subatomic particles that make possible all the electronic miracles in our living room.

The problem is that these two theories cannot be united into a single comprehensive one. They are quite dissimilar, based on different assumptions, different mathematics, and different physical pictures.

If a unified field theory were possible, the energy at which unification would take place is the Planck energy. This is the point at which Einstein's theory of gravity breaks down completely. It is the energy of the Big Bang and the energy at the center of a black hole.

The Planck energy is 10^{19} billion electron volts, which is a quadrillion times the energy produced by the Large Hadron Collider at CERN, the most powerful particle accelerator on Earth.

At first, it would seem hopeless to probe the Planck energy, since it is so enormous. But a Type III civilization, which has more than 10^{20} times more energy than a Type I civilization, has enough power to do so. So a Type III civilization may be able to play with the fabric of space-time and bend it at will.

They may reach this incredible energy scale by creating a particle accelerator much bigger than the Large Hadron Collider. The LHC is a circular tube in the shape of a doughnut seventeen miles in circumference, surrounded by huge magnetic fields.

When a stream of protons is injected into the LHC, the magnetic fields bend their path into a circle. Then pulses of energy are periodically sent into the doughnut, causing them to accelerate. There are two beams of protons traveling inside the tube in opposite directions. When they reach maximum velocity, they collide head-on, unleashing the energy of fourteen trillion electron volts, the largest burst of energy ever created artificially. (This collision is so powerful that some people have worried that perhaps it might open up a black hole that could consume the Earth. This is not a valid concern. In fact, there are naturally occurring subatomic particles that hit the Earth all the time with energies much larger than fourteen trillion electron volts. Mother Nature can hit us with cosmic rays far more powerful than the puny ones created in our labs.)

BEYOND THE LHC

The LHC has made many headlines, including the discovery of the elusive Higgs boson, which won the Nobel Prize for two physicists, Peter Higgs and Francois Englert. One of the main purposes of the LHC was to complete the last piece of the puzzle, called the Standard Model of particles, which is the most advanced version of the quantum theory and gives us a complete description of the universe at low energies.

The Standard Model is sometimes called "the theory of almost everything" because it accurately describes the low-energy universe that we see around us. But it cannot be the final theory, for several reasons:

1. It makes no mention of gravity. Worse, when we combine the Standard Model with Einstein's theory of gravity, the hybrid theory blows up, giving us nonsense (calculations become infinite, meaning that the theory is useless).

2. It has a strange collection of particles that seem quite contrived. It has thirty-six quarks and anti-quarks, a series of

Yang-Mills gluons, leptons (electrons and muons), and Higgs bosons.

3. It has nineteen or so free parameters (masses and couplings of particles) that have to be put in by hand. These masses and couplings are not determined by the theory; no one knows why they have these numerical values.

It's hard to believe that the Standard Model, with its motley collection of subatomic particles, is nature's final theory. It's like taking Scotch tape and wrapping up a platypus, aardvark, and whale and calling it Mother Nature's finest creation, the end product of millions of years of evolution.

The next big particle accelerator currently in the planning stage is the International Linear Collider (ILC), consisting of a straight tube approximately thirty miles long in which beams of electrons and anti-electrons will collide. The current plan is that it will be based in the Kitakami Mountains of Japan and is expected to cost roughly $20 billion, of which half will be supplied by the Japanese government.

Although the maximum energy of the ILC will be only one trillion electron volts, in many ways it will be superior to the LHC. When smashing protons into each other, the collision is extremely difficult to analyze because the proton has a complicated structure. It contains three quarks, held together by particles called "gluons." The electron, however, has no known structure. It looks like a point particle. Therefore, when an electron collides with an anti-electron, it is a clean, simple interaction.

Even with these advances in physics, our Type 0 civilization cannot directly probe the Planck energy. But this is within the realm of a Type III civilization. Building accelerators like the ILC may be a crucial step in being able to one day test how stable space-time is and determine whether we might be able to take shortcuts through it.

ACCELERATOR IN THE ASTEROID BELT

Eventually, an advanced civilization might build a particle accelerator the size of the asteroid belt. A circular beam of protons would be sent around the belt, guided by gigantic magnets. On Earth, particles are sent inside a large circular tube containing a vacuum. But since the vacuum of outer space is better than any vacuum on the Earth, this accelerator does not need a tube at all.

All it needs is a series of gigantic magnetic stations placed strategically around the belt, making a circular path for the proton beam. It is somewhat like a relay race. Each time the protons go past a station, a surge of electrical energy powers the magnets, which kick the proton beam so that it moves to the next station at the correct angle. Each time the proton beam passes by a magnetic station, more energy is pumped into the beam in the form of laser power, until it gradually reaches the Planck energy.

Once the accelerator attains this energy, it can focus that energy onto a single point. A wormhole should open up there. It would then be injected with enough negative energy to stabilize it so it doesn't collapse.

What might a trip through the wormhole look like? No one knows, but an educated guess was made by the physicist Kip Thorne of Caltech when he helped to advise the directors of the film **Interstellar.** Thorne used a computer program to trace the paths of light beams as they went past one, so that you could get a visual feeling for what this trip might look like. Unlike the usual cinematic representations, this was the most rigorous attempt yet to visualize this journey on film.

(In the movie, as you approach a black hole, you see a gigantic black sphere, called the event horizon. As you go through the event horizon, you pass the point of no return. Inside the black sphere lies the black hole itself, a tiny point of incredible density and gravity.)

In addition to building gigantic particle accelerators, there are a few other ways that physicists have considered exploiting wormholes. One possibility is that the Big Bang was so explosive that it might have inflated tiny wormholes that existed in the infant universe 13.8 billion years ago. When the universe began to expand exponentially, these wormholes may have expanded with it. This means that, although at present no one has ever seen one, they might be a naturally occurring phenomenon. Some physicists have speculated about how to go about finding one

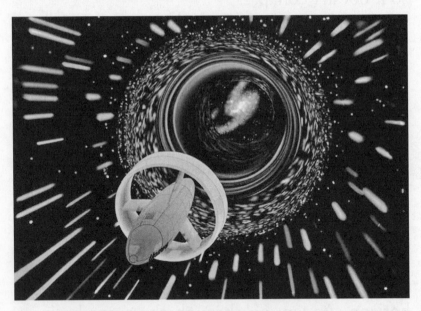

As a starship enters a wormhole, it must withstand intense radiation due to quantum fluctuations.
In principle, only string theory has the ability to calculate the fluctuations, so you can determine if you will survive.

in space. (To find a naturally occurring wormhole, which is the subject of several **Star Trek** episodes, one would look for an object that distorts the passage of starlight in a particular way, perhaps so it resembles a sphere or a ring.)

Another possibility, also explored by Kip Thorne and his collaborators, is to find a tiny one in the vacuum and then expand it. Our latest understanding of space is that it may be frothing with tiny wormholes as universes spring into existence and then vanish again. So if you had enough energy, you might be able to manipulate a preexisting wormhole and inflate it.

There is one problem, however, with all these proposals. The wormhole is surrounded by particles of gravity, called gravitons. As you are about to pass through it you will encounter quantum corrections in the form of gravitational radiation. Normally, quantum corrections are small and can be ignored. But calculations show that these corrections are infinite as you pass through a wormhole, so the radiation would likely be lethal. Also the radiation levels are so strong that the wormhole may close, making a passage impossible. There is a debate among physicists today about how dangerous it might be to travel through a wormhole.

Einstein's relativity is no longer of any use as we enter the wormhole. Quantum effects are so large that we need a higher theory to take us through. Currently the only one capable of doing this is string theory, which is one of the strangest ever proposed in physics.

QUANTUM FUZZINESS

What theory can unify general relativity and the quantum theory at the Planck energy? Einstein spent the last thirty years of his life chasing after a "theory of everything" that could allow him to "read the mind of God," but he failed. This remains one of the biggest questions facing modern physics. The solution will reveal some of the most important secrets of the universe, and, using it, we may be able to explore time travel, wormholes, higher dimensions, parallel universes, even what happened before the Big Bang. Furthermore, the answer will determine whether or not humanity can travel the universe at faster-than-light velocities.

To understand this, we have to understand the basis of the quantum theory, the Heisenberg uncertainty principle. This innocent sounding principle states that no matter how sensitive your instruments, you can never know both the velocity and position of any subatomic particle, say an electron. There is always a quantum "fuzziness." Thus, a startling picture emerges. An electron is actually a collection of different states, with each state describing an electron in a different position with a different velocity. (Einstein hated this principle. He believed in "objective reality," which is the commonsense notion that objects exist in definite, well-defined states and that you can determine the exact position and velocity of any particle.)

But quantum theory states otherwise. When you look in a mirror, you are not seeing yourself as you really are. You are made up of a vast collection of waves. So the image you see in the mirror is actually an average, a composite of all these waves. There is even a small probability that some of these waves can spread out all over your room and into space. In fact, some of your waves can even spread out to Mars or beyond. (One problem we give our Ph.D. students is to calculate the probability that some of your waves spread out to Mars and that one day you will get out of bed and wake up on the Red Planet.)

These waves are called "quantum corrections" or "quantum fluctuations." Normally, these corrections are small, so the commonsense notion is perfectly fine, since we are a collection of atoms and can only see averages. But at the subatomic level, these quantum corrections can be large, so that electrons can be several places at the same time and exist in parallel states. (Newton would be shocked if you explained to him how the electrons in transistors can exist in parallel states. These corrections make modern electronics possible. So if we could somehow turn off this quantum fuzziness, all of these marvels of technology would stop functioning and society would be thrown almost a hundred years into the past, before the electric age.)

Fortunately, physicists can calculate these quantum corrections for subatomic particles and make predictions for them, some of which are valid to incredible accuracy, to one part in ten trillion. In fact, the quantum theory is so accurate that it is perhaps the most successful theory of all time. Nothing else can match its accuracy when applied to ordinary matter. It may be the most bizarre theory ever proposed in history (Einstein once said that the more successful the quantum theory becomes, the stranger it becomes), but it has one small thing going for it: it is undeniably correct.

So the Heisenberg uncertainty principle forces us to reevaluate what we know about reality. One result is that black holes cannot really be black. Quantum theory says that there must be quantum corrections to pure blackness, so black holes are actually gray. (And they emit a faint radiation called Hawking radiation.) Many textbooks say that at the center of a black hole, or at the beginning of time, there is a "singularity," a point of infinite gravity. But infinite gravity violates the uncertainty principle. (In other words, there is no such thing as a "singularity"; it is simply a word we invent to disguise our ignorance about what occurs when the equations don't work out. In the quantum theory, there are no singularities because there is a fuzziness that prevents knowing the precise location of the black hole.) Similarly, it is often stated that a pure vacuum is a state of pure nothingness. The concept of "zero" violates the uncertainty principle, so there is no such thing as pure nothingness. (Instead, the vacuum is a cauldron of virtual matter and antimatter particles constantly springing in and out of existence.) And there is no such thing as absolute zero, the temperature at which all motion stops. (Even as we approach it, atoms continue to move slightly, which is called the zero-point energy.)

When we try to formulate a quantum theory of gravity, a problem occurs, however. The quantum corrections to Einstein's theory are described by particles we call "gravitons." Just like a photon is a particle of light, a graviton is a particle of gravity. Gravitons are so elusive that they have never been seen in the laboratory. But physicists are confident that they do exist, since they are essential to any quantum theory of gravity. When we try to calculate with these gravitons, however, we find that quantum corrections are infinite. Quantum gravity is riddled with corrections that blow up the equations. Some of the greatest minds in physics have tried to solve this problem, but all have failed.

So this is one goal of modern physics: to create a quantum theory of gravity where the quantum corrections are finite and calculable. In other words, Einstein's theory of gravity allows for the formation of wormholes, which may one day give us shortcuts through the galaxy. But Einstein's theory cannot tell us if these wormholes are stable or not. To calculate these quantum corrections, we need a theory that combines relativity with the quantum theory.

STRING THEORY

So far, the leading (and only) candidate to solve this problem is something called string theory, which says that all matter and energy in the universe is composed of tiny strings. Each vibration of the string corresponds to a different subatomic particle. So the electron is not really a point particle. If you had a supermicroscope, you would see that it is not a particle at all but a vibrating string. The electron appears to be a point particle only because the string is so tiny.

If the string vibrates at a different frequency, it corresponds to a different particle, such as a quark, mu meson, neutrino, photon, and so on. That is why physicists have discovered such a ridiculous number of subatomic particles. There are literally hundreds, all because they are just different vibrations of a tiny string. In this way, string theory can explain the quantum theory of subatomic particles. According to string theory, as the string moves, it forces space-time to curl up exactly as Einstein predicted, and hence it unifies Einstein's theory and the quantum theory in a very pleasing fashion.

This means that subatomic particles are just like musical notes. The universe is a symphony of strings, physics represents the harmonies of these notes, and the "mind of God" that Einstein chased after for so many decades is cosmic music resonating through hyperspace.

So how does string theory banish the quantum corrections that have bedeviled physicists for decades? String theory possesses something called "supersymmetry." For every particle, there is a partner: a superparticle or "sparticle." For example, the partner of the electron is the "selectron." The partner of the quark is the "squark." So we have two types of quantum corrections, those coming from ordinary particles and those from the sparticles. The beauty of string theory is that the quantum corrections coming from these two sets of particles exactly cancel each other out.

Thus, string theory gives us a simple but elegant way to eliminate these infinite quantum corrections. They vanish because the theory reveals a new symmetry that gives the theory its mathematical power and its beauty.

To artists, beauty may be an ethereal quantity that they aspire to capture in their works. But to a theoretical physicist, beauty is symmetry. It is also an absolute necessity when probing the ultimate nature of space and time. For example, if I have a snowflake and rotate it by 60 degrees, the snowflake remains the same. In the same way, a kaleidoscope creates beautiful patterns because it uses mirrors to repeatedly duplicate an image so it fills up 360 degrees. We say that the snowflake and kaleidoscope both possess radial symmetry; that is, they remain the same after a certain radial rotation.

Let's say I have an equation containing many subatomic particles and then I shuffle or rearrange them among one another. If the equation remains the same after interchanging these particles, I would then say that the equation has a symmetry.

THE POWER OF SYMMETRY

Symmetry is not just a matter of aesthetics. It is a powerful way to eliminate imperfections and anomalies in your equations. If you rotate the snowflake, you can rapidly spot any defects by comparing the rotated version with the original. If they are not the same, then you have a problem that needs correcting.

In the same way, when constructing a quantum equation, we often find that a theory is infested with tiny anomalies and divergences. But if the equation has a symmetry, then these defects are eliminated. In the same way, supersymmetry takes care of the infinities and imperfections often found in a quantum theory.

As a bonus, it turns out that supersymmetry is the largest symmetry ever found in physics. Supersymmetry can take all known subatomic particles and mix them together or rearrange them while preserving the original equation. In fact, supersymmetry is so powerful that it can take Einstein's theory, including the graviton and the subatomic particles of the Standard Model, and rotate them or interchange them. This gives us a pleasing and natural way to unify Einstein's theory of gravity and subatomic particles.

String theory is like a gigantic cosmic snowflake, except that each prong of the snowflake represents the entire set of Einstein's equations and the Standard Model of subatomic particles. So each prong of the snowflake represents all the particles of the universe. As we rotate the snowflake, all the particles of the universe are interchanged. Some physicists have noted that even if Einstein had never been born, and billions of dollars were never spent on smashing atoms to create the Standard Model, then all of twentieth-century physics might have been discovered if you simply possessed string theory.

Most important, supersymmetry cancels the quantum corrections of particles with those of sparticles, leaving us with a finite theory of gravity. That is the miracle of string theory. This also explains the answer to the question most often heard about string theory: Why does it exist in ten dimensions? Why not thirteen, or twenty?

This is because the number of particles in string theory can vary with the dimensionality of space-time. In higher dimensions, we have more particles, since there are more ways in which particles can vibrate. When we try to cancel the quantum corrections from the particles against the corrections from the sparticles, we find that this cancellation can happen only in ten dimensions.

Usually, mathematicians create new, imaginative structures that physicists later incorporate into their theories. For example, the theory of curved surfaces was worked out by mathematicians in the nineteenth century and was later incorporated into Einstein's theory of gravity in 1915. But this time, the reverse happened. String theory has opened up so many new branches of mathematics that the mathematicians were startled. Young, aspiring mathematicians, who usually scorn applications of their discipline, have to learn string theory if they want to be on the cutting edge.

Although Einstein's theory allows for the possibility of wormholes and faster-than-light travel, you need string theory to calculate how stable these wormholes are in the presence of quantum corrections.

In summary, these quantum corrections are infinite, so removing these infinities is one of the fundamental problems in physics. String theory eliminates these quantum corrections, because it has two types of quantum corrections that precisely cancel each other. This precise cancellation between particles and sparticles is due to supersymmetry.

However, as elegant and powerful as string theory is, it is not enough; it must ultimately face the final challenge, which is experiment.

CRITICISMS OF STRING THEORY

Although this picture is compelling and persuasive, there are valid criticisms one can make of the theory. First, since the energy at which string theory (or any theory of everything for that matter) unifies all of physics is the Planck energy, no machine on Earth is powerful enough to rigorously test it. A direct test would involve creating a baby universe in the laboratory, which is obviously out of the question given current technology.

Second, like any physical theory, it has more than one solution. For example, Maxwell's equations, which govern light, have an infinite number of solutions. This is not a problem because, at the very beginning of any experiment, we specify what we are studying, whether it's a light bulb, laser, or a TV. Then later, given these initial conditions, we solve the equations of Maxwell. But if we have a theory of the universe, then what are its initial conditions? Physicists believe that a "theory of everything" should dictate its own initial state, that is, they would prefer that the initial conditions of the Big Bang somehow emerge from the theory itself. String theory, however, does not tell you which of its many solutions is the correct one for our universe. And, without initial conditions, string theory contains an infinite number of parallel universes, called the multiverse, each one as valid as the next. So we have an embarrassment of riches, with string theory predicting not only our own familiar universe but perhaps an infinite number of other equally valid alien universes as well.

Third, perhaps the most startling prediction of string theory is that the universe is not four-dimensional at all but exists in ten dimensions. In all of physics, nowhere have we seen a prediction this bizarre, a theory of space-time that selects out its own dimensionality. This was so strange that many physicists at first dismissed it as science fiction. (When string theory was first proposed, the fact that it could only exist in ten dimensions was a source of ridicule. Nobel laureate Richard Feynman, for example, would tease John Schwarz, one of the founders of string theory, by asking him, "So John, how many dimensions are we in today?")

LIVING IN HYPERSPACE

We know that any object in our universe can be described by three numbers: length, width, and height. If we add time, then four numbers can describe any event in the universe. For example, if I want to meet someone in New York City, I might say that we should meet at Forty-Second Street and Fifth Avenue, on the tenth floor, at noon. But to a mathematician the need for only three or four coordinates might seem arbitrary, since there is nothing special about three or four dimensions. Why should the most fundamental feature of the physical universe be described by such ordinary numbers?

So mathematicians have no problem with string theory. But to visualize these higher dimensions, physicists often use analogies. When I was a child, I used to spend many hours gazing at the Japanese Tea Garden in San Francisco. Watching the fish swim in the shallow pond, I asked myself a question that only a child would ask: "What would it be like to be a fish?" What a strange world they would see, I thought. They would think the universe was only two-dimensional. They could only swim in this limited space by moving sideways, but never up or down. Any fish who dared mention a third dimension beyond the pond would be considered a crackpot. I then imagined there was a fish living in the pond who would scoff anytime someone mentioned hyperspace, since the universe was just what you could touch and feel, nothing more. Then I imagined grabbing that fish and lifting him into the world of "up." What would he see? He would see beings moving without fins. A new law of physics. Beings breathing without water. A new law of biology. Then I imagined putting the scientist fish back into the pond and he would have to explain to the other fish the incredible creatures that live in the world of "up."

Similarly, perhaps we are the fish. If string theory is proven correct, it means that there are unseen dimensions beyond our familiar four-dimensional world. But where are these higher dimensions? One possibility is that six of the ten original dimensions have "curled up" so they cannot be seen anymore. Think of taking a sheet of paper and rolling it up into a tight tube. The original sheet was two-dimensional, but the rolling-up process has created a one-dimensional tube. From a distance, you only see the one-dimensional tube, but in reality it is still two-dimensional.

In the same way, string theory says that the universe was originally ten-dimensional, but for some reason six of these dimensions curled up, leaving us with the illusion that our world has only four. Although this feature of string theory seems fantastic, efforts are under way to actually measure these higher dimensions.

But how do higher dimensions help string theory unify relativity and quantum mechanics? If you try to unify the gravitational, nuclear, and electromagnetic forces into a single theory, you find that there is not enough "room" in four dimensions to do this. They are like pieces of a jigsaw puzzle that don't fit together. But once you start to add more and more dimensions, you find enough room to assemble these lower theories, like matching jigsaw pieces together to make the whole.

For example, think of a two-dimensional world of Flatlanders, who, like cookie men, can only move left or right, but never "up." Imagine that there was once a beautiful three-dimensional crystal that exploded, showering fragments onto Flatland. Over the years, the Flatlanders have re-assembled this crystal into two large fragments. But as hard as they try, they are unable to fit these last two fragments together. Then one day, a Flatlander makes the outrageous proposal that if they move one fragment "up," into the unseen third dimension, then the two fragments would fit together and form a beautiful three-dimensional crystal. So the key to re-creating the crystal was moving the fragments through the third dimension. By analogy, these two fragments are relativity theory and the quantum theory, the crystal is string theory, and the explosion was the Big Bang.

Even though string theory fits the data neatly, we still need to test it. Although as discussed a direct test is not possible, most physics is done indirectly. For example, we know that the sun is made mainly of hydrogen and helium, yet no one has ever visited the sun. We know the sun's composition because we analyze it indirectly, looking at sunlight through a prism, which breaks it up into bands of colors. By studying these bands within the rainbow, we can identify the fingerprint of hydrogen and helium. (In fact, helium was not found on Earth first. In 1868, scientists discovered evidence of a strange new element when analyzing sunlight during an eclipse, which was christened "helium," meaning "metal from the sun." It wasn't until 1895 that direct evidence of helium was discovered on the Earth, when scientists realized it was a gas and not a metal.)

DARK MATTER AND STRINGS

In the same way, string theory might be proven via a variety of indirect tests. Since each vibration of the string corresponds to a particle, we can in our particle accelerators search for entirely new particles that represent higher "octaves" of the string. The hope is that by smashing protons together at trillions of volts, you briefly create a new particle among the debris that is predicted by string theory. This, in turn, may help explain one of the great unsolved problems in astronomy.

In the 1960s, when astronomers examined the rotation of the Milky Way galaxy, they found something strange. It was rotating so fast that, by Newton's laws, it should fly apart, yet the galaxy has been stable for about ten billion years. In fact, the galaxy rotated about ten times faster than it should according to traditional Newtonian mechanics.

This posed a tremendous problem. Either Newton's equations were wrong (which was almost unthinkable) or there was an invisible halo of unknown matter surrounding the galaxies, increasing their mass sufficiently for gravity to hold them together. This meant that perhaps the pictures we see of gorgeous galaxies with their beautiful spiral arms are incomplete, that they are actually surrounded by a gigantic invisible halo that is ten times more massive than the visible galaxy. Since photographs of galaxies only show the beautiful swirling mass of stars, whatever is holding the mass together must not interact with light—it must be invisible.

Astrophysicists dubbed this missing mass "dark matter." Its existence forced them to revise their theories, which said that the universe is made mainly of atoms. We now have maps of dark matter throughout the universe. Although it is invisible, it bends starlight just as anything with mass should. Therefore, by analyzing the distortion of starlight surrounding galaxies, we can use computers to calculate the presence of dark matter and map its distribution across the universe. Sure enough, this map shows that most of the total mass of a galaxy exists in this form.

In addition to being invisible, dark matter has gravity, but you can't hold it in your hand. Since it does not interact with atoms at all (because it is electrically neutral) it will pass through your hand, the floor, and through the crust of the Earth. It would oscillate between New York and Australia as if the Earth did not exist at all, except that it would be bound by Earth's gravity. So although dark matter is invisible, it still interacts via gravity with other particles.

One theory is that dark matter is a higher vibration of the superstring. The leading candidate is the superpartner of the photon, which is called the "photino," or "little photon." It has all the right properties to be dark matter: it is invisible because it does not interact with light, and yet it has weight and is stable.

There are several ways to prove this conjecture. The first is to create dark matter directly with the Large Hadron Collider by smashing protons into each other. For a brief instant of time, a particle of dark matter would be formed inside the accelerator. If this is possible, it would have enormous repercussions for science. It would represent the first time in history that a new form of matter has been found that is not based on atoms. If the

LHC is not powerful enough to produce dark matter, then perhaps the ILC can.

There also is another way to prove this conjecture. The Earth is moving in a wind of this invisible dark matter. The hope is that a dark matter particle may smash into a proton inside a particle detector, creating a shower of subatomic particles that might be photographed. At present, there are physicists around the world patiently waiting to find the signature of a collision between matter and dark matter in their detectors. There is a Nobel Prize waiting for the first physicist to do so.

If dark matter is found, either with particle accelerators or with ground-based sensors, we will be able to compare its properties with those predicted by string theory. In this way, we will have evidence to evaluate the validity of the theory.

Although finding dark matter would be a great step toward proving string theory, other proofs are possible. For example, Newton's law of gravity governs the motion of large objects like stars and planets, but little is known about the force of gravity acting over small distances, like a few inches or feet. Since string theory postulates higher dimensions, this means that Newton's famous inverse square law (that gravity diminishes in proportion with the square of the distance) should be violated at small distances because

Newton's law is predicated on three dimensions. (If space were four-dimensional, for instance, then gravity should diminish in proportion to the inverse cube of the distance. So far, tests of Newton's law of gravity have not shown any evidence of a higher dimension, but physicists aren't giving up.)

Another possible avenue is to send gravity wave detectors into space. The Laser Interferometer Gravitational-Wave Observatory (LIGO) based in Louisiana and Washington State was successful in picking up gravity waves from colliding black holes in 2016 and colliding neutron stars in 2017. A modified version of the space-based Laser Interferometer Space Antenna (LISA) may be able to detect gravity waves from the instant of the Big Bang. The hope is that one might be able to "run the videotape backward" and make conjectures about the nature of the pre–Big Bang era. This would allow a crude test of some of the predictions of string theory concerning the pre–Big Bang universe.

STRING THEORY AND WORMHOLES

Still other tests of string theory may involve finding other exotic particles predicted by the theory, such as micro black holes, which resemble subatomic particles.

We have seen how physics allows us to speculate about civilizations far into the future, making reasonable conjectures based on their energy consumption. Civilizations can be expected to evolve from a Type I planetary civilization to a Type II stellar civilization and finally to a Type III galactic civilization. A galactic civilization, in turn, is likely to explore the galaxy via von Neumann probes or by laser porting their consciousness across the galaxy. The key point is that a Type III civilization may be able to access the Planck energy, the point where space-time becomes unstable and faster-than-light travel might be possible. But to calculate the physics of faster-than-light travel, we need a theory that goes beyond Einstein's theory, which might well be string theory.

The hope is that using string theory, we will be able to calculate the quantum corrections necessary to analyze exotic phenomena such as time travel, interdimensional travel, wormholes, and what happened before the Big Bang. For example, assume that a Type III civilization is capable of manipulating black holes and thereby creating a gateway to a parallel universe through a wormhole. Without string theory, it is impossible to calculate what happens when you enter. Will it explode? Will gravitational radiation close it just as you enter it? Will you be able to pass through it and live to tell about it?

String theory should be capable of calculating how much gravitational radiation you would encounter when you pass through the wormhole and answer these questions.

Another hotly debated question among physicists is what happens if you enter a wormhole and go backward in time. If you then kill your grandfather before you are born, then you have a paradox. How can you exist at all if you just killed your ancestor? Einstein's theory actually allows for time travel (if negative energy exists) but says nothing about how to resolve these paradoxes. String theory, because it is a finite theory in which everything can be calculated, should be able to resolve all these mind-twisting paradoxes. (My own strictly personal opinion is that the river of time forks into two rivers when you enter a time machine—in other words, the timeline splits. This means that you have killed someone else's grandfather who looks just like your own grandfather but exists in another timeline in an alternate universe. So the multiverse resolves all time paradoxes.)

At present, however, because of the complexity of the mathematics of string theory, physicists have not been able to apply it to these questions. This is a mathematical problem, not an experimental one, so perhaps one day an enterprising physicist will be able to definitively calculate the properties of wormholes and hyperspace. Instead of idly speculating about faster-than-light travel, a physicist using string theory has the ability to determine whether this might be possible. But we will have to wait until the theory is sufficiently understood to make this determination.

END OF THE DIASPORA?

So there is a possibility that a Type III civilization may be able to use a quantum theory of gravity to achieve faster-than-light-speed spaceships.

But what are the implications of this for humanity?

Earlier, we noted that a Type II civilization, bound by the speed of light, may establish space colonies that eventually branch off, creating many distinct genetic lineages which may eventually lose all contact with the mother planet.

The question remains, What happens when a Type III civilization masters the Planck energy and begins to make contact with these branches of humanity?

History may repeat itself. For example, the Great Diaspora ended with the coming of the airplane and modern technology, giving us a rapid international transportation network. Today, we can take a short plane trip over continents that once took our ancestors tens of thousands of years to cross.

In the same way, when we make the transition from a Type II civilization to a Type III civilization, we will, by definition, have enough power to explore the Planck energy, the point at which space-time becomes unstable.

If we assume that this makes faster-than-light travel possible, it means that a Type III civilization might be able to unify the various Type II colonies that have spread out across the galaxy. Given our common human heritage, it may make possible the creation of a new galactic civilization, as envisioned by Asimov.

As we have seen earlier, the amount of genetic divergence that humanity may experience over several tens of thousands of years in the future is roughly the same as the divergence that has already occurred since the Great Diaspora. The key point is that we have maintained our humanity throughout. A young child, born in one culture, can easily grow up and mature in another totally different culture, even if the two cultures may be separated by a vast cultural chasm.

This also means that Type III archeologists, curious about ancient human migrations, may try to retrace the ancient migration routes of various branches of Type II civilizations across the galaxy. Galactic archeologists may look for signs of various ancient Type II civilizations.

In the Foundation saga, our heroes are in search

of the ancestral planet that gave birth to the Galactic Empire, whose name and location were lost in the chaos of galactic prehistory. Given that the human population numbers in the trillions, with millions of inhabited planets, this seems like a hopeless task. But by exploring the most ancient planets in the galaxy, they find ruins of the earliest planetary colonies. They see how planets were abandoned because of wars, disease, and other calamities.

Likewise, a Type III civilization may emerge from a Type II civilization and try to retrace the various branches that were explored centuries earlier by sub-light-speed spaceships. In the same way that our current civilization is enriched by the presence of so many different types of cultures, each with a different history and perspective, a Type III civilization may be enriched by interacting with the many divergent civilizations that emerged during a Type II civilization.

So the creation of faster-than-light spaceships may make the dream of Asimov come true, unifying humanity into one galactic civilization.

As Sir Martin Rees has said, "If humans avoid self-destruction, the post-human era beckons. Life from Earth could spread through the entire galaxy, evolving into a teeming complexity far beyond what we can even conceive. If so, our tiny planet—this pale blue dot floating in space—could be the most important place in the entire Galaxy. The first interstellar voyagers from Earth would have a mission that would resonate through the entire Galaxy and beyond."

But eventually any advanced civilization will have to face the ultimate challenge to their existence, which is the end of the universe itself. We have to ask the question, Can an advanced civilization, with all its vast technology, evade the death of everything there is? Perhaps the only hope for intelligent life is to evolve into a Type IV civilization.

Some say the world will end in fire,
Some say in ice.
From what I've tasted of desire
I hold with those who favor fire.
—ROBERT FROST, 1920

Eternity is an awful long time—
especially towards the end.
—WOODY ALLEN

14 LEAVING THE UNIVERSE

The Earth is dying.

In the movie **Interstellar,** a strange blight has hit the planet, causing crops to fail and agriculture to collapse. People are starving. Civilization is slowly crumbling as it faces a devastating famine.

Matthew McConaughey plays a former NASA astronaut who is given a dangerous mission. Earlier, a wormhole mysteriously opened up near Saturn. It is a gateway that will transport anyone who enters it to a distant part of the galaxy, where there might be new inhabitable worlds. Desperate to save humanity, he volunteers to enter the wormhole and search for a new home for humanity among the stars.

Meanwhile, back on Earth, scientists are desperately trying to find the secret of the wormhole. Who made it? And why did it appear just as humanity was about to perish?

Slowly, the truth dawns on the scientists. The technology to make this wormhole is millions of years more advanced than ours. The beings who made it are actually our descendants. The creators are so advanced that they live in hyperspace, beyond our familiar universe. They have built a gateway to the past, to send advanced technology to save their ancestors (us). By saving humanity, they will actually save themselves. According to Kip Thorne, who in addition to being a physicist was one of the producers of the film, the inspiration for the physics behind the movie comes from string theory.

If we survive, one day we will face a similar crisis, except this time, the universe is dying.

One day in the far future, the universe will go cold and dark; stars will cease to shine as the universe is plunged into a Big Freeze. All life will cease to exist when the universe itself dies, eventually reaching near absolute zero in temperature.

But the question is, Are there any loopholes? Can we avoid this cosmic doom? Can we, like Matthew McConaughey, find salvation in hyperspace?

In order to understand how the universe might die, it is important to analyze the predictions of the far future given to us by Einstein's theory of gravity and then analyze the startling new revelations that have been made in the last decade.

According to these equations, there are three possibilities for the ultimate fate of the universe.

BIG CRUNCH, BIG FREEZE, OR BIG RIP

The first is the Big Crunch, when the expansion of the universe slows down, stops, and reverses itself. In this scenario, the galaxies in the heavens will eventually halt and begin to contract. Temperatures will rise dramatically as the distant stars come closer and closer. Eventually, all the stars coalesce into a primordial superheated mass. In some scenarios, there might even be a Big Bounce and the Big Bang could start all over again.

The second is the Big Freeze, when the growth of the universe continues unabated. The second law of thermodynamics states that total entropy always increases, so eventually, the universe will grow cold as matter and heat become more diffuse. The stars will cease to shine, the night sky would become totally black, and temperatures would plunge to near absolute zero, when even molecules cease almost all their motion.

For decades, astronomers have been trying to determine which scenario determines the fate of our universe. This is done by calculating its average density. If the universe is dense enough, then there is enough matter and gravity to attract the distant galaxies and reverse the expansion, so that the Big Crunch becomes a realistic possibility. If the universe lacks sufficient mass, then there is not enough gravity to reverse the expansion and the universe goes into a Big Freeze. The critical density separating these two scenarios is roughly six hydrogen atoms per cubic meter.

But in 2011, the Nobel Prize in physics was given to Saul Perlmutter, Adam Riess, and Brian Schmidt for a discovery that overturned decades of cherished belief. They found that the universe, instead of slowing down in its expansion, was actually speeding up. The universe is 13.8 billion years old, but about 5 billion years ago, it began to accelerate exponentially. Today, the universe is expanding in a runaway fashion. **Scientific American** claimed, "The astrophysical community was stunned to learn that the universe was driving itself apart." These astronomers came to this astounding conclusion by analyzing super-nova explosions in distant galaxies to determine the rate at which the universe expanded billions of years ago. (One type of supernova explosion, called Type Ia, has a fixed luminosity, so we can accurately measure its distance using its brightness. If one has a headlamp of known luminosity it's easy to tell how far away it is, but if you don't know its brightness it's difficult to tell its distance. A headlamp of known brightness is a "standard candle." A Type Ia supernova acts as a standard candle, so it's easy to tell its distance.) When analyzing these supernovae, scientists found that they were moving away from us, just as expected. But to their shock, they found that closer supernovae appeared to be moving away more rapidly than they should, indicating the rate of expansion was accelerating.

So in addition to the Big Freeze and Big Crunch, a third alternative began to emerge from the data, the Big Rip, which is like the Big Freeze on steroids. It is a vastly accelerated time frame for the life cycle of the universe.

In the Big Rip, the distant galaxies eventually move away from us so fast that they exceed the speed of light and disappear from view. (This does not violate special relativity, because it is space that is expanding faster than light. Material objects cannot move faster than light, but empty space can stretch and expand at any speed.) This means that the night sky will become black, because light from the distant galaxies is moving away so quickly it can't reach us.

Eventually, this exponential expansion becomes so great that not only is the galaxy torn apart, the solar system is ripped apart, and the very atoms making up our bodies are also torn apart. Matter as we know it cannot exist in the final stages of the Big Rip.

Scientific American writes, "Galaxies would be destroyed, the solar system would unbind and eventually all the planets would burst asunder as the rapid expansion of space rips apart its very atoms. Finally, our universe would end in an explosion, a singularity of literally infinite energy."

Bertrand Russell, the great British philosopher and mathematician, once wrote:

All the devotion, all the inspiration, all the noonday brightness of human genius, are destined to extinction in the vast death of the solar system, and [the] whole temple of man's achievement must inevitably be buried beneath the debris of a universe in ruins . . . Only within the scaffolding of these truths, only on the firm foundation of unyielding despair, can the soul's habitation henceforth be safely built.

Russell wrote about "a universe in ruins" and "unyielding despair" in response to predictions by physicists of the Earth's eventual demise. But he did not foresee the coming of the space program. He did not foresee that advances in technology might allow us to escape the death of our planet.

But although we might one day avoid the death of the sun with our spaceships, how will we avoid the death of the universe itself?

FIRE OR ICE?

The ancients, in some sense, anticipated many of these violent scenarios.

Every religion, it seems, has some mythology to explain the birth and death of the universe.

In Norse mythology, the Twilight of the Gods is called Ragnarok, the day of reckoning, when the world is blanketed in unending snow and ice and the heavens freeze over. The world witnesses the final battle between the frost giants and the Norse gods of Asgard. In Christian mythology, we have Armageddon, when the forces of good and evil clash for the last time. The Four Horsemen of the Apocalypse appear, foretelling the Final Judgment. In Hindu mythology, there is no final end of days at all. Instead, there is an unending series of cycles, each lasting about eight billion years.

But after thousands of years of speculation and wonder, science is beginning to understand how our world will evolve and eventually die.

For the Earth, the future lies in fire. In five or so billion years we will have the last nice day on our home planet, then the sun will exhaust its hydrogen fuel and expand into a red giant star. Eventually the sun will set the sky on fire. The oceans will boil and the mountains will melt. The Earth will be engulfed by the sun, and will orbit like a burnt-out cinder within its fiery atmosphere. There is a biblical reference that says, from ashes to ashes, dust to dust. Physicists say, from stardust we came, to stardust we will return.

The sun itself will suffer a different fate. After the red giant phase, it will eventually exhaust all its nuclear fuel, shrink, and go cold. It will become a small white dwarf star, about the size of the Earth, and eventually die as a dark dwarf star, a piece of nuclear waste drifting in the galaxy.

Unlike our sun the Milky Way galaxy will die in fire. About four billion years from now, it will collide with Andromeda, the nearest spiral galaxy. Andromeda is roughly twice the size of the Milky Way, so it will be a hostile takeover. Computer simulations of the collision show that the two galaxies will enter a death dance as they orbit around each other. Andromeda will rip off many of the arms of the Milky Way, dismembering it. The black holes at the center of both galaxies will orbit around each other and finally collide, merging into a bigger black hole, and a new galaxy will emerge from the collision, a giant elliptical galaxy.

In each of these scenarios, it is important to realize that rebirth is also part of this cosmic cycle. Planets, stars, and galaxies get recycled. Our sun, for example, is probably a third-generation star. Each time a star explodes, the dust and gas it spews into space reseed the next generation of stars.

Science also gives us an understanding of the life of the entire universe. Until recently, astronomers thought they understood its history and ultimate fate trillions of years into the future. They had speculated that it is evolving slowly in five epochs:

1. In the first epoch, the first billion years after the Big Bang, the universe was filled with hot opaque clouds of ionic molecules, too hot for electrons and protons to condense into atoms.
2. In the second epoch, a billion years after the Big Bang, the universe cooled down enough so that atoms, stars, and galaxies could emerge from the chaos. Empty space suddenly became crystal clear, and stars lit

up the universe for the first time. We are living in this era now.

3. In the third epoch, about one hundred billion years after the Big Bang, the stars will have exhausted most of their nuclear fuel. The universe will consist mainly of small red dwarf stars, which burn so slowly that they can shine for trillions of years.

4. In the fourth epoch, trillions of years after the Big Bang, all the stars will finally burn out and the universe will go completely black. Only dead neutron stars and black holes remain.

5. In the fifth epoch, even black holes begin to evaporate and disintegrate, so the universe becomes a sea of nuclear waste and drifting subatomic particles.

With the discovery of the accelerating universe, this entire scenario might be compressed into billions of years. The Big Rip upsets the entire applecart.

DARK ENERGY

What is causing this sudden change in our understanding of the ultimate fate of the universe?

According to Einstein's theory of relativity, there are two sources of energy that drive the evolution of the universe. The first is the curvature of space-time, which creates the familiar gravity fields surrounding the stars and galaxies. This curvature is what keeps our feet on the ground. This is the energy source most studied by astrophysicists.

But there is also a second source of power, which is usually ignored. It is the energy of nothingness, the energy of the vacuum, called dark energy (not to be confused with dark matter). The very emptiness of space contains energy.

The most recent calculations show that this dark energy acts like antigravity and it is pushing the universe apart. The more the universe expands, the more dark energy there is, which causes it to expand even faster.

At present, the best data indicate that about 69 percent of the matter/energy (since matter and energy are interchangeable) in the universe is contained in dark energy. (By contrast, dark matter makes up about 26 percent, atoms of hydrogen and helium make up about 5 percent, and higher elements, which make up the Earth and our own bodies, only make up a tiny 0.5 percent.) So dark energy, which is pushing the galaxies away from us, is clearly the dominant force in the universe, much larger than the energy contained in the curvature of space-time.

One of the central problems in all of cosmology is therefore to understand the origin of dark energy. Where does it come from? Will it ultimately destroy the universe?

Usually, when we simply combine relativity and the quantum theory in a crude shotgun marriage, we can get a prediction for dark energy, but the resulting prediction is off by a factor of 10^{120}, which is the largest mismatch in the history of science. Nowhere do we find a discrepancy this large. It indicates that something is terribly wrong with our understanding of the universe. So the unified field theory, instead of being a scientific curiosity, becomes essential to understanding how everything works. The solution to this question will tell us the fate of the universe and all intelligent creatures in it.

ESCAPE FROM APOCALYPSE

Given that the fate of the universe is likely to die a cold death in the distant future, what can we do about it? Can these cosmic forces be reversed?

There are at least three options.

The first is to do nothing and let the life cycle of the universe play out. As it gets colder and colder, intelligent beings will adjust and think slower and slower, according to physicist Freeman Dyson. Eventually, a simple thought may take millions of years, but these beings will never notice because all other beings will think slower as well. It would be possible to have intelligent conversations between these beings, even if it takes millions of years. So from this point of view, everything would seem normal.

Living in such a cold world may actually be quite interesting. Quantum leaps, which are extremely unlikely in a human life span, may begin to occur routinely. Wormholes may open up and close before our eyes. Bubble universes may pop into and out of existence. These beings may see them all the time because their brains operate so slowly.

However, this is only a temporary solution, because eventually molecular motion will become so slow that information cannot be transferred from one place to another. At this point all activity, including thinking, no matter how slow, will cease. One desperate hope is that the acceleration caused by dark energy will suddenly disappear all by itself before this happens. Since no one knows why the universe is accelerating, there is that possibility.

BECOMING TYPE IV

In the same vein, the second option is that we evolve into a Type IV civilization and learn to utilize energy beyond our own galaxy. I once gave a talk on cosmology and discussed the Kardashev scale. Afterward a ten-year-old boy came up to me and said I was wrong. There must be a Type IV civilization, beyond the usual Type I, II, III of the Kardashev classification. I corrected him and told him that there were only planets, stars, and galaxies in the universe and that hence a Type IV civilization is impossible. There was no energy source beyond the galaxy.

Later, I realized that perhaps I was too impatient with the boy.

Remember that each type of civilization is ten to one hundred billion times more powerful than the previous type. Since there are about one hundred billion galaxies in the visible universe, a Type IV civilization could harness the energy of the entire visible universe.

Perhaps the extragalactic energy source is dark energy, which is by far the largest source of matter/energy in the universe. How might a Type IV civilization manipulate dark energy and reverse the Big Rip?

Because, by definition, a Type IV civilization can harness extragalactic energy, they might manipulate some of the extra dimensions revealed by string theory and create a sphere in which dark energy reverses polarity, so that the cosmic expansion is reversed. Outside the sphere, the universe might still be expanding exponentially. But inside the sphere, the galaxies evolve normally. In this way, a Type IV civilization could survive even if the universe is dying all around it.

In some sense, it would act like a Dyson sphere. But although the purposes of the Dyson sphere would be to trap sunlight inside, the purpose of this sphere would be to trap dark energy, so that the expansion could be contained.

The final possibility is to create a wormhole through space and time. If the universe is dying, then one option might be to leave it and enter another, younger one.

The original picture given to us by Einstein is that the universe is a huge expanding bubble. We live on the skin of the bubble. The new picture given to us by string theory indicates that there are other bubbles out there, each one a solution of the string equations. In fact, there is a bubble bath of universes, creating a multiverse.

Many of these bubbles are microscopic and pop into existence in a mini Big Bang and then rapidly collapse. Most of them are of no consequence to us, since they live out their short lives in the vacuum of space. Stephen Hawking has called this constant churning of universes in the vacuum the "space-time foam." So nothingness is not empty but is full of universes in constant motion. Strangely, this means that even within our bodies there are vibrations within the space-time foam, but they are so tiny that we are blissfully unaware of them.

The startling aspect of this theory is that if the Big Bang happened once, it can happen again and again. So a new picture emerges of baby universes budding from mother universes, and our universe is nothing but a tiny patch of a much larger multiverse.

(Occasionally, a tiny fraction of these bubbles do not vanish back into the vacuum but expand enormously due to dark energy. This is perhaps the origin of our own universe, or our universe may be the result of the collision of two bubbles or the fissioning of a bubble into smaller bubbles.)

As we saw in the last chapter, an advanced civilization might be able to build a gigantic particle accelerator the size of the asteroid belt that could open up a wormhole. If it is stabilized by negative energy, then it might provide an escape route to another universe. We've already discussed using the Casimir effect to create this negative energy. But another source of negative energy is these higher dimensions. They may serve two purposes: they may change the value of dark energy, thereby preventing the Big Rip, or they may create negative energy to help stabilize a wormhole.

Each bubble or universe in the multiverse has different laws of physics. Ideally, we want to enter a parallel universe where atoms are stable (so our bodies do not disintegrate when we enter it) but the amount of dark energy is much lower, so that it expands enough to cool down and allow habitable planets to form but not so much that it accelerates into an early Big Freeze.

INFLATION

All these speculations at first seem preposterous, but the latest cosmological data from our satellites seem to support this picture. Even the skeptics are forced to admit that the multiverse idea is consistent with the theory called "inflation," which is a supercharged version of the old Big Bang theory. In this scenario, just before the Big Bang, there was an explosion called inflation that created the universe in the first 10^{-33} seconds, much faster than the original theory. This idea, originally proposed by Alan Guth of MIT and Andrei Linde of Stanford, solved a number of cosmological mysteries. For example, the universe seems much flatter and more uniform than predicted by Einstein's theory. But if the universe underwent a cosmic expansion, it would flatten out, much like inflating an enormous balloon. The surface of the inflated balloon seems flat because of its size.

Also, when we look in one direction of the

universe and then look 180 degrees in the opposite direction, we see that the universe is pretty much the same no matter where we look. This requires some form of mixing between its different parts, but because light has a finite velocity, there is simply not enough time for information to travel across these vast distances. Hence, the universe should look lumpy and disorganized because there was not enough time to mix the matter. Inflation solves this by postulating that, at the beginning of time, the universe was a tiny patch of uniform matter. As inflation expanded this patch, it created what we see today. And because inflation is a quantum theory, there is a small but finite probability that it can happen again.

Although the inflation theory has had undeniable success in explaining the data, there is still a debate among cosmologists as to the underlying theory behind it. There is considerable evidence from our satellites that shows that the universe underwent a rapid inflation, but precisely what drove this inflation is not known. So far, the leading way to explain inflation theory is through string theory.

I once asked Dr. Guth if it might be possible to create a baby universe in the laboratory. He replied that he actually did the calculation. One would have to concentrate a fantastic amount of heat at one point. If the baby universe were to be formed inside a lab, it would explode violently in a Big Bang. However, it would explode in another dimension, so, from our point of view, the baby universe would vanish. However, we would still feel the shock wave of it being born, which would be equivalent to the explosion of many nuclear weapons. So, he concluded, if we did create one, we would have to run quickly!

NIRVANA

The multiverse can also be viewed from the perspective of theology, where all religions fall into two categories: religions in which there was an instant of creation, and religions that are eternal. For example, the Judeo-Christian philosophy talks about a creation, a cosmic event when the universe was born. (Not surprisingly, the original calculations of the Big Bang were done by a Catholic priest and physicist, Georges Lemaître, who believed that Einstein's theory was compatible with Genesis.) However, in Buddhism, there is no god at all. The universe is timeless, with no beginning or end. There is only Nirvana. These two philosophies seem totally in opposition to each other. Either the universe had a beginning or it didn't.

But a melding of these two diametrically opposed philosophies is possible if we adopt the multiverse concept. In string theory, our universe did in fact have a cataclysmic origin, the Big Bang. But we live in a multiverse of bubble universes. These bubble universes, in turn, are floating in a much larger arena, a ten-dimensional hyperspace, which had no beginning.

So Genesis is happening all the time within the larger arena of Nirvana (hyperspace).

This then gives us a simple and elegant unification of the Judeo/Christian origin story with Buddhism. Our universe did in fact have a fiery beginning, but we coexist in a timeless Nirvana of parallel universes.

STAR MAKER

This takes us back to the work of Olaf Stapledon, who imagined that there is a Star Maker, a cosmic being that creates and discards entire universes. He is like a celestial painter, constantly conjuring up new universes, tinkering with their properties, and then moving on to the next one. Each universe has different laws of nature and different life-forms.

The Star Maker himself was outside these universes and could see all of them in their totality as he painted on the canvas of the multiverse. Stapledon writes, "Each cosmos . . . was itself gifted with its own peculiar time, in such a manner that the whole sequence of events with any single cosmos could be viewed by the Star Maker not only from within the cosmical time itself but also externally, from the time proper to his own life, with all the cosmical epochs co-existing together."

This is very similar to the way in which string theorists view the multiverse. Each universe in the multiverse is a solution of the string equations, each with its own laws of physics, each with its own time scales and units of measurement. As Stapledon said, one must be outside of normal time, outside of all these universes, to see these bubbles all at once.

(This also is reminiscent of the way Saint Augustine viewed the nature of time. If God was all-powerful, then He could not be bound by earthly concerns. In other words, divine beings do not have to rush to meet deadlines or make appointments. In some sense, therefore, God must be outside of time. In the same way, the Star Maker and string theorists, gazing at the bubble bath of universes in the multiverse, are also outside of time.)

But if we have a bubble bath of possible universes, then which one is ours? This raises the question of whether our universe was designed by a higher being or not.

When we examine the forces of the universe, we find that it seems to be "tuned" just right to make intelligent life possible. For example, if the nuclear force were a bit stronger, the sun would have burned out millions of years ago. If it were a bit weaker, the sun would never have ignited in the first place. The same applies to gravity. If it were a bit stronger, we would have had a Big Crunch billions of years ago. If it were a bit weaker, we would have had a Big Freeze instead. In both cases, the nuclear and gravitational forces are "tuned" just right to make intelligent life on Earth possible. When we examine other forces and parameters, we find the same pattern.

Several philosophies have emerged to address the problem of the narrow range of these fundamental laws that could allow life.

The first is the Copernican principle, which simply states that there is nothing special about the Earth. So the Earth is just a piece of cosmic dust wandering aimlessly through the cosmos. It is just a coincidence that the forces of nature are "tuned" just right.

The second is the anthropic principle, which states that our very existence places enormous constraints on what kinds of universes can exist. A weak form of this principle simply says that the laws of nature should be such that life is possible, since we exist and are contemplating those laws. Any universe is just as good as any other, but only our universe has intelligent beings who can ponder and write about this. But a much stronger version states that it is so unlikely that intelligent life exists that perhaps the universe is compelled in some way to allow intelligent life to exist, that perhaps the universe was designed for it to be so.

The Copernican principle says that our universe is not special, while the anthropic principle says that it is. Strangely, while both principles are diametrically opposite of each other, they are both compatible with the universe as we know it.

(When I was in second grade, I clearly remember my teacher explaining this idea to me. She said that God so loved the Earth that He put it just the right distance from the sun. If it was too close, the oceans would boil. If it was too far, the oceans would freeze. So God chose the Earth to be just right from the sun. This was the first time I had ever heard a scientific principle explained in this way.)

The way to resolve this problem without invoking religion is the existence of exoplanets, most of which are too close or too far from the sun to support life. We are here today because of luck. It was luck that we live in the Goldilocks zone around the sun.

Likewise, the explanation for why the universe seems to be fine-tuned to allow for life as we know it is because of luck, because there are billions of parallel universes that are not fine-tuned for life, that are completely lifeless. We are the lucky ones who can live to tell about it. So the universe is not necessarily designed by a superior being. We are here to discuss the question because we live in a universe compatible with life.

But there is another way to look at this problem. This is the philosophy that I prefer and the one that I am working on at present. In this approach, there are many universes in the multiverse, but most are not stable and will eventually decay down to a more stable universe. Many other universes might have existed in the past, but they didn't last and were subsumed into ours. In this picture, our universe survives because it is one of the most stable.

So my point of view combines both the Copernican and anthropic principle. I believe that our universe is not special, as in the Copernican principle, except for two features: that it is very stable and that it is compatible with life as we know it. So instead of having an infinite number of parallel universes floating in the Nirvana of hyperspace, most of them are unstable, and perhaps only a handful of them survive to create life like ours.

The final word on string theory has yet to be written. Once the full theory is solved, we can compare it with the amount of dark matter in the universe and the parameters describing subatomic particles, which may settle the question of whether the theory is correct or not. If it is correct, string theory may also explain the mystery of dark energy, which physicists believe is the engine that may one day destroy the universe. And if we are fortunate enough to evolve into a Type IV civilization, capable of harnessing extragalactic power, then string theory may explain how the death of the universe itself may be avoided.

Perhaps some enterprising young mind, reading this book, will be inspired to complete the last chapter in the history of string theory and answer the question of whether the death of the universe can be reversed.

THE LAST QUESTION

Isaac Asimov once said that of all the short stories he had written, his favorite was "The Last Question," which gave a startling new vision of life trillions of years into the future and explained how humanity might confront the end of the universe.

In that story, people have asked over the aeons

whether the universe must necessarily die or whether it was possible to reverse the expansion and prevent the universe from freezing over. When asked, "Can entropy ever be reversed?" the master computer replies each time, "There is insufficient data for a meaningful answer."

Finally, in the far future trillions of years from now, humanity has outgrown the confines of matter itself. Humans have evolved into pure energy beings that can transport themselves across the galaxy. Without the shackles of matter, they can visit the far reaches of the galaxy as pure consciousness. Their physical bodies are immortal but stored in some distant, forgotten solar system, so that their minds are free to roam. But each time they ask the fateful question, "Can entropy be reversed?" they get the same response: "There is insufficient data for a meaningful answer."

Finally, the master computer is so powerful that it cannot be placed on any planet and is housed in hyperspace. The trillions of minds that make up humanity fuse with it. As the universe enters its final death throes, the computer finally solves the problem of reversing entropy. Just as the universe dies, the master computer declares "Let there be light!" And there was light.

So ultimately, the future of humanity is to evolve into a god that can create an entirely new universe and begin again. This was a masterful work of fiction. But let us now analyze this short story from the point of modern physics.

As we mentioned in the last chapter, we might be able to laser-port our consciousness at the speed of light within the next century or so. Eventually, laser porting may become a vast intergalactic superhighway, carrying billions of minds racing across the galaxy. So Asimov's vision of beings of pure energy exploring the galaxy is not such a far-fetched idea.

Next, the master computer becomes so large and powerful that it has to be placed in hyperspace, and eventually humanity merges with it. Maybe someday we can become like the Star Maker and from our vantage point in hyperspace look down and see our universe, coexisting with other universes in the multiverse, each containing billions of galaxies. Analyzing the landscape of possible universes, we may choose a new universe that is still young, that can provide a new home. We would choose a universe that has stable matter, like atoms, and is young enough that stars can create new solar systems to spawn new forms of life. So the distant future, instead of being a dead end for intelligent life, might see the birth of a new home for it. If this is the case, then the death of the universe is not the end of the story.

Our only chance of long-term survival is
not to remain lurking on planet Earth, but
to reach out into space . . . But I am an
optimist. If we can avoid disaster for the next
two centuries, our species should be safe,
as we spread into space. Once we establish
independent colonies, our entire future
should be safe.
—STEPHEN HAWKING

Every dream begins with a dreamer. Always
remember, you have within you the strength,
and the passion, to reach for the stars to
change the world.
—HARRIET TUBMAN

NOTES

PROLOGUE

1   **One day about seventy-five thousand years
     ago:** A. R. Templeton, "Genetics and Recent
     Human Evolution," **International Journal of
     Organic Evolution** 61, no. 7 (2007): 1507–19.
     See also **Supervolcano: The Catastrophic Event
     That Changed the Course of Human History;
     Could Yellowstone Be Next?** (New York:
     MacMillan, 2015).

2   **Stark evidence of this cataclysm:** Although
     there is universal agreement that the eruption of

the supervolcano at Toba was a truly catastrophic event, it should be pointed out that not all scientists believe it altered the direction of human evolution. One group, from Oxford University, analyzed sediments in Lake Malawi in Africa going back tens of thousands of years into the past. By drilling into the lake bottom, one can retrieve sediments that were deposited in the ancient past and hence re-create ancient weather conditions. Analysis of this data from the time of the Toba volcano showed no significant sign of permanent climate change, which casts doubt on the theory. However, it remains to be seen if this result can be generalized to other areas besides Lake Malawi. Another theory is that the bottleneck in human evolution about seventy-five thousand years ago was caused by slow environmental effects rather than a sudden collapse of the environment. Further research is required to definitively settle the question.

CHAPTER 1: PREPARING FOR LIFTOFF

28 **As a youth, he spent most of his time:** Newton's three laws of motion are:
- An object in motion stays in motion, unless acted on by an outside force. (This means that our space probes can reach the distant planets with minimal fuel once they are in space, because they basically coast their way to the planets, since there is no friction in space.)
- Force equals mass times acceleration. This

is the fundamental law behind Newtonian mechanics, which makes possible the building of skyscrapers, bridges, and factories. At any university, a first-year course in physics is basically solving this equation for different mechanical systems.

- For every action, there is an equal and opposite reaction. This is the reason why rockets can move in outer space.

These laws work perfectly well when shooting space probes throughout the solar system. However, they inevitably break down in several important domains: (a) extremely fast velocities approaching the speed of light, (b) extremely intense gravitational fields, such as near a black hole, and (c) extremely small distances found inside the atom. To explain these phenomena, we need Einstein's theory of relativity and also the quantum theory.

29 **"To place one's feet on the soil of asteroids":** Chris Impey, **Beyond** (New York: W.W. Norton, 2015).

32 **"That Professor Goddard":** Impey, **Beyond**, p. 30.

33 **Wernher von Braun would take the sketches, dreams, and models:** Historians still debate precisely how much cross-fertilization there was between pioneers like Tsiolkovsky, Goddard, and von Braun. Some claim that each worked in near total isolation and independently rediscovered one another's work. Others claim that there was considerable interaction between them, especially

because much of their work was published. But it is known that the Nazis made inquiries to Goddard, asking for his advice. So it is safe to say that von Braun, because he had access to the German government, was fully aware of the developments of his predecessors.

35   **"I plan on traveling to the Moon":** Hans Fricke, **Der Fisch, der aus der Urzweit kam** (Munich: Deutscher Taschenbuch-Verlag, 2010), pp. 23–24.

39   **"I reach for the stars, but sometimes I hit London":** See Lance Morrow, "The Moon and the Clones," **Time,** August 3, 1998. For more on the political legacy of von Braun, see M. J. Neufeld, **Wernher von Braun: Dreamer of Space, Engineer of War** (New York: Vintage, 2008). Also, parts of this discussion were based on a radio interview I conducted with Mr. Neufeld in September 2007. Many have written about this great scientist, who opened up the space age but did it using financial backing from the Nazis, and have come to differing conclusions.

40   **While the U.S. rocketry program proceeded by fits and starts:** See R. Hal and D. J. Sayler, **The Rocket Men: Vostok and Voskhod, the First Soviet Manned Spaceflights** (New York: Springer Verlag, 2001).

50   **"Congress came to see NASA primarily as a jobs program":** See Gregory Benford and James Benford, **Starship Century** (New York: Lucky Bat Books, 2014), p. 3.

CHAPTER 2: NEW GOLDEN AGE OF SPACE TRAVEL

65 **"The whole idea is to preserve the Earth":** Peter Whoriskey, "For Jeff Bezos, The Post Represents a New Frontier," **Washington Post,** August 12, 2013.

66 **In the 1990s, an unexpected discovery caught scientists by surprise:** See R. A. Kerr, "How Wet the Moon? Just Damp Enough to Be Interesting," Science Magazine 330 (2010): 434.

69 **The Chinese have announced that they will put their astronauts on the moon:** See B. Harvey, China's Space Program: From Conception to Manned Spaceflight (Dordrecht: Springer-Verlag, 2004).

70 **One factor that limits how long our astronauts can stay on the moon:** See J. Weppler, V. Sabathier, and A. Bander, "Costs of an International Lunar Base" (Washington, D.C.: Center for Strategic and International Studies, 2009); https://csis.org/publication/costs -international-lunar-base.

CHAPTER 3: MINING THE HEAVENS

90 **Planetary Resources estimates that the platinum:** See www.planetaryresources.com.

CHAPTER 4: MARS OR BUST

101 **"Failure is an option here [at SpaceX]":** For more quotes from Elon Musk, see www.investo

pedia.com/university/elon-musk-biography/elon
-musk-most-influential-quotes.asp.

101 **"They say Mars is the new black"**: See https://
manofmetropolis.com/nick-graham-fall-2017
-review.

102 **"I really don't have any other motivation"**: The
Guardian, September 2016; www.theguardian
.com/technology/2016/sep/27/elon-musk-spacex
-mars-exploration-space-science.

102 **"I'm convinced"**: The Verge, October 5, 2016;
www.theverge.com/2016/10/5/13178056/
boeing-ceo-mars-colony-rocket-space-elon-musk.

103 **"I think it's good for there to be multiple
paths to Mars"**: Business Insider, October 6,
2016; www.businessinsider.com/boeing-spacex
-mars-elon-musk-2016-10.

104 **"NASA applauds all those"**: Ibid.

110 **Bill Gerstenmaier, of NASA's Human
Exploration and Operations Directorate:** See
www.nasa.gov/feature/deep-space-gateway-to
-open-opportunities-for-distant-destinations.

CHAPTER 5: MARS: THE GARDEN PLANET

132 **"Actually, it was Sputnik"**: Interview on *Science
Fantastic* radio, June 2017.

135 **Another outlandish attempt to form an
isolated colony:** See R. Reider, *Dreaming the
Biosphere* (Albuquerque: University of New
Mexico Press, 2010).

CHAPTER 6: GAS GIANTS, COMETS, AND BEYOND

162 **Using Newton's laws, astronomers can calculate:** The calculation of the Roche limit and tidal forces requires only an elementary application of Newton's law of gravity. Because a moon is a spherical object and not a point particle, the force of attraction from a gas giant like Jupiter is larger on the side facing Jupiter than the gravity on the far side. This causes the moon to bulge a bit. But one can also calculate the force of gravity, which holds the moon together via its own gravitational pull. If the moon gets close enough, the force of gravity that is pulling the moon apart balances the force of gravity that holds the moon together. At that point, the moon begins to disintegrate. This gives us the Roche limit. All the rings of the gas giants that have been documented lie within the Roche limit. This indicates, but does not prove, that the rings of the gas giants were caused by tidal forces.

166 **Beyond the gas giants, at the outer reaches of our solar system:** Comets from the Kuiper Belt and Oort Cloud probably have different origins. Originally, the sun was a gigantic ball of hydrogen gas and dust, perhaps a few light-years across. As the gas began to collapse because of gravity, it began to spin faster. At that point, some of the gas collapsed into a spinning disk, which eventually condensed into the solar system. Since this spinning disk contained water, this

created a ring of comets in the outer reaches of the solar system. This became the Kuiper Belt. However, some of the gas and dust did not condense into this spinning disk. Some of it condensed into chunks of stationary ice, roughly tracing out the original outlines of the original protostar. This became the Oort Cloud.

CHAPTER 7: ROBOTS IN SPACE

179 **"AlphaGo can't even play chess":** Discover Magazine, April 2017; discovermagazine. com/2017/april-2017/cultivating-common-sense.

199 **In 2017, a controversy arose between two billionaires:** Many fear that AI could revolutionize the job market, putting millions of people out of work. This may very well happen, but there are other trends that might reverse this effect. New jobs will open up—in designing, repairing, maintaining, and servicing robots— as the industry explodes in size, perhaps rivaling the automobile industry. Furthermore, there are many classes of jobs that cannot be replaced by robots for decades to come. For example, semiskilled, nonrepetitive workers—such as janitors, police, construction workers, plumbers, gardeners, contractors, et cetera—cannot be replaced by robots. Robots, for example, are too primitive to pick up garbage. In general, jobs that will be difficult to automate with robots include jobs involving (a) common sense, (b) pattern recognition, and (c) human

interactions. For example, in a law firm, the paralegal might be replaced, but lawyers are still needed to argue cases before a live jury or judge. Middlemen, especially, may find themselves out of work, so they will have to add value to their services (i.e., intellectual capital). This means adding analysis, experience, intuition, and innovation, which robots are deficient in.

199 **"We are ourselves creating our own successors":** Samuel Butler, **Darwin Among the Machines;** www.historyofinformation.com/expanded.php?id=3849.

200 **"I visualize a time when we will be to robots what dogs are to humans":** For more quotes from Claude Shannon, see www.quotes-inspirationa.com/quote/visualize-time-robots-dogs-humans-121.

201 **"It is ridiculous to talk about such things so early":** Raffi Khatchadourian, "The Doomsday Invention," **New Yorker,** November 23, 2015; www.newyorker.com/magazine/2015/11/23/doomsday-invention-artificial-intelligence-nick-bostrom.

201 **When addressing the Zuckerberg/Musk controversy:** The debate about the dangers and benefits of AI has to be put into perspective. Every discovery can be used for good or evil. When the bow and arrow was first invented, it was mainly used to hunt small game, like squirrels and rabbits. But eventually, it evolved into a formidable weapon that could be used to hunt other humans. Similarly, when the first

airplanes were invented, they were used for recreation and delivering the mail. But eventually, they evolved into weapons that can deliver bombs. Similarly, AI for many decades to come will be a useful invention that can generate jobs, new industries, and prosperity. But eventually, these machines can pose an existential risk if they become too intelligent. At what point will they become dangerous? I personally believe that tipping point will occur when they become self-aware. Currently, robots do not know they are robots, but that could change radically in the future. However, this tipping, in my opinion, probably won't be reached until near the end of this century, giving us time to prepare.

202 **He believes that by 2045, we will reach the "singularity":** One should be careful when analyzing one aspect of the singularity: that future generations of robots can be smarter than the previous generation so that one can rapidly create superintelligent robots. One can, of course, create computers that have increasingly large amounts of memory, but does this mean that they are "smarter"? In fact, no one has been able to demonstrate even a single computer that can create a second-generation computer that is more intelligent. There is, in fact, no rigorous definition of the word **smart**. This does not mean that it is impossible for this to happen, it only means that the process is ill defined. In fact, it is not clear how this will be accomplished.

208 **In order to create self-aware machines:** The

key to human intelligence, in my opinion, is our ability to simulate the future. Humans constantly plan, scheme, daydream, ponder, and muse about the future. We can't help it. We are prediction machines. But one of the keys to simulating the future is understanding the laws of common sense, of which there are billions. These laws, in turn, depend on understanding the basic biology, chemistry, and physics of the world around us. The more accurate our understanding of these laws, the more accurate our simulation of the future will be. At present, the common sense problem is one of the major hurdles in AI. Massive attempts to codify all the laws of common sense have all failed. Even a child has more common sense than our most advanced computer. So in other words, a robot that tries to take over the world from humans will fail miserably because it doesn't understand the simplest things about our world. It is not enough for a robot to try to dominate humans; one has to master the simplest laws of common sense in order to carry out a plan. For example, giving a robot the simple goal of robbing a bank will ultimately result in failure because the robot cannot realistically map out all possible future scenarios.

CHAPTER 8: BUILDING A STARSHIP

224 **In a subsequent phase of the project:** R. L. Forward, "Roundtrip Interstellar Travel Using

Laser-Pushed Lightsails," **Journal of Spacecraft** 21, no. 2 (1984): 187–95.

225 **Laser-propelled nanoships:** See G. Vulpetti, L. Johnson, and L. Matloff, **Solar Sails: A Novel Approach to Interplanetary Flight** (New York: Springer, 2008).

226 **"There will some day appear velocities far greater than these":** Jules Verne, **From the Earth to the Moon.** Quoted at www.space.com/5581 -nasa-deploy-solar-sail-summer.html.

233 **The idea was developed by nuclear physicist Ted Taylor:** G. Dyson, **Project Orion: The True Story of the Atomic Spaceship** (New York: Henry Holt, 2002).

237 **There are several ways in which to release the power of fusion peacefully:** S. Lee and S. H. Saw, "Nuclear Fusion Energy—Mankind's Giant Step Forward," **Journal of Fusion Energy** 29, 2, 2010.

240 **The nuclear fusion rocket is conceptually sound:** The fundamental reason why magnetic fusion has not yet been attained on the Earth is because of the stability problem. In nature, giant balls of gas can be compressed so that the star ignites, because gravity compresses the gas uniformly. However, magnetism involves two poles, north and south. Therefore, it is impossible to compress gas uniformly using magnetism. When you squeeze gas magnetically in one area, it bulges out the other end. (Think of trying to squeeze a balloon. If you pinch the balloon in one place, it expands in another.) One idea is

to create a doughnut-shaped magnetic field and have the gas compressed on the inside of the doughnut. But physicists have failed to compress hot gas for more than a tenth of a second, which is too brief to create a self-sustaining fusion reaction.

240 **They would utilize the greatest energy source in the universe:** Although antimatter rockets convert matter into energy with 100 percent efficiency, there are also some hidden losses. For example, some of the energy of a matter/antimatter collision is in the form of neutrinos, which cannot be harvested to create usable energy. Our bodies are continually radiated by neutrinos from the sun, yet we feel nothing. Even when the sun sets, our bodies are irradiated by neutrinos that have gone right through the planet Earth. In fact, if you could somehow shine a beam of neutrinos through solid lead, it might penetrate a light-year of lead before it is finally stopped. So the neutrino energy created by matter/antimatter collisions is lost and cannot be used to generate power.

245 **The ramjet fusion rocket is another enticing concept:** R. W. Bussard, "Galactic Matter and Interstellar Flight," **Astronautics Acta** 6 (1960): 179–94.

249 **Space elevators would be a game-changing application:** D. B. Smitherman Jr., "Space Elevators: An Advanced Earth-Space Infrastructure for the New Millennium," NASA pub. CP 2000-210429.

250   **"Probably about fifty years after everyone stops laughing"**: NASA Science, "Audacious and Outrageous: Space Elevators"; https://science .nasa.gov/science-news/science-at-nasa/2000/ ast07sep_1.

253   **One day, a boy read a children's book and changed world history**: Einstein's theory of special relativity is based on the simple sentence: "The speed of light is constant in any inertial frame [i.e. in any uniformly moving frame]." This violates Newton's laws, which say nothing about the speed of light. In order for this law to be satisfied, there have to be vast changes in our understanding of the laws of motion. From that one statement, one can show that:

- The faster you move in a rocket ship, the slower time beats inside that rocket.
- Space is compressed within that rocket the faster you move.
- You get heavier the faster you move.

As a result, this means that at the speed of light, time would stop and you would become infinitely flat and infinitely heavy, which is impossible. Hence, you cannot break the light barrier. (In the Big Bang, however, the universe expanded so rapidly that the expansion exceeded the speed of light. This is not a problem, because it's empty space that is stretching faster than light. Material objects, however, are forbidden to go faster than light.)

The only way known to go faster than light is to invoke Einstein's general theory of relativity, where space-time becomes a fabric that can stretch, bend, and even tear. The first way is via "multiply connected spaces" (wormholes), in which two universes are joined together like Siamese twins. If we take two parallel sheets of paper and then punch a hole that connects them, then this gives us a wormhole. Or, you could somehow compress space in front of you, so that you can hop over the compressed space and travel faster than light.

258 **Although physicists have seen no evidence of negative matter:** Stephen Hawking proved a powerful theorem, which states that negative energy is essential to any solution of Einstein's equations that allows for time travel or wormhole starships.

Negative energy is not allowed under ordinary Newtonian mechanics. However, negative energy is allowed by the quantum theory via the Casimir effect. It has been measured in the laboratory and found to be extremely tiny. If we have two large parallel metal plates, then the Casimir energy is proportional to the inverse distance of separation of the plates raised to the third power. In other words, negative energy rapidly increases in energy as the two plates are brought together.

The problem is that these plates have to be brought together to within subatomic distances, which is not possible with today's technology. We have to assume that a very advanced civilization has somehow mastered the ability to harness vast amounts of negative energy to make time machines and wormhole spaceships possible.

259 **I once interviewed the Mexican theoretical physicist Miguel Alcubierre:** See M. Alcubierre, "The Warp Drive: Hyperfast Travel Within General Relativity," **Classical and Quantum Gravity** 11, no. 5 (1994): L73–L77. When I interviewed Alcubierre for the Discovery Channel, he was confident that his solution of Einstein's equations was a significant contribution, but he was wary of the difficulties it faced if one actually tried to build a warp drive engine. First, the space-time inside the warp bubble was causally separate from the outside world. This meant that it was impossible to steer the starship or direct it from the outside. Second, and most important, it required vast amounts of negative matter (which has never been found) and negative energy (which only exists in minute quantities). So, he concluded, major hurdles have to be solved before a workable warp engine can be built.

CHAPTER 9: KEPLER AND A UNIVERSE OF PLANETS

267 **Bruno, Galileo's predecessor:** William Boulting, **Giordano Bruno: His Life, Thought, and Martyrdom** (Victoria, Australia: Leopold Classic Library, 2014).

268 **"This space we declare to be infinite":** Ibid.

274 **A big breakthrough came with the 2009**

launch of the Kepler spacecraft: For more on the Kepler spacecraft, see the NASA website: http://www.kepler.arc.nasa.gov.

The Kepler spacecraft focused on one tiny spot in the Milky Way galaxy. Even then, it has found evidence of four thousand or so planets orbiting other stars. But from that tiny spot, we can extrapolate to the entire galaxy and hence get a rough analysis of the planets in the Milky Way. Succeeding missions after the Kepler will focus on different regions of the Milky Way galaxy, hoping to find different types of extrasolar planets, and more Earth-like ones.

275 **"There are planets out there that have no counterpart in our solar system":** Interview with Professor Sara Seager, **Science Fantastic** radio, June 2017.

278 **"This is a game changer in exoplanetary science":** Christopher Crockett, "Year In Review: A Planet Lurks Around the Star Next Door," **Science News,** December 14, 2016.

278 **"It's absolutely phenomenal":** Interview with Professor Sara Seager, **Science Fantastic** radio, June 2017.

281 **"This is an amazing planetary system":** See www.quotes.euronews.com/people/michael -gillion-KAp4OyeA.

CHAPTER 10: IMMORTALITY

302 **Yet another proposal to colonize the galaxy is to send embryos:** A. Crow, J. Hunt, and A. Hein, "Embryo Space Colonization to Overcome

the Interstellar Time Distance Bottleneck,"
Journal of the British Interplanetary Society 65
(2012): 283–85.

308 **"Every sign, including genetics, says there's
some causality":** Linda Marsa, "What It Takes to
Reach 100," **Discover Magazine,** October 2016.

312 **The mechanism of aging is slowly being
revealed:** It is sometimes said that immortality
violates the second law of thermodynamics,
which indicates that everything, including living
organisms, will eventually decay, rot, and die.
However, there is a loophole in the second law,
which states that (in a closed system) entropy
(disorder) will inevitably increase. The key
word is **closed.** If you have an open system
(where energy can be added from the outside),
then entropy can be reversed. This is how a
refrigerator works. The motor at the bottom of
the refrigerator pushes gas through a pipe, which
causes the gas to expand, causing the refrigerator
to cool down. When applied to living things,
it means that entropy can be reversed as long
as energy is added from the outside (which is
sunlight).

So our very existence is possible because
sunlight can energize plants, and we can
consume these plants and use this energy to
repair the damage caused by entropy. Hence,
we can reverse entropy locally. When discussing
human immortality, one can therefore evade the
second law by adding new energy locally from
the outside (such as in the form of changes in

diet, exercise, gene therapy, absorbing new types of enzymes, et cetera).

315 **"I don't think the time is quite right, but it's close":** Quoted in Michio Kaku, **The Physics of the Future** (New York: Anchor Books, 2012), p. 118.

316 **What happens if we solve the problem of aging?:** The point here is that, in the main, all the pessimistic predictions of population collapse made back in the 1960s failed to materialize. In fact, the rate of expansion of the world population is actually slowing down. But the point is that the absolute population of the world is still increasing, especially in sub-Saharan Africa, so it is difficult to actually estimate the world population in 2050 and 2100. Some demographers, however, have claimed that, if trends continue, ultimately the world population could flatten out and become stable. If so, then the world population could reach a plateau of some sort and hence avoid a population catastrophe. But this is still conjectural.

322 **"I'm as fond of my body as anyone":** See https://quotefancy.com/quote/1583084/Danny -Hillis-I-m-as-fond-of-my-body-as-anyone-but -if-I-can-be-200-with-a-body-of-silicon.

CHAPTER 11: TRANSHUMANISM AND TECHNOLOGY

346 **"It just completely changes the landscape":** Andrew Pollack, "A Powerful New Way to Edit

DNA," **New York Times,** March 3, 2014; www
.nytimes.com/2014/03/04/health/a-powerful
-new-way-to-edit-DNA.html.

349 **"No one really has the guts to say it":** See
Michio Kaku, **Visions** (New York: Anchor Books,
1998), p. 220 and Michio Kaku, **The Physics of
the Future,** p. 118.

350 **"My prediction is that by the year 2100":**
Kaku, **The Physics of the Future,** p. 118.

352 **Francis Fukuyama of Stanford has warned:**
F. Fukuyama, "The World's Most Dangerous
Ideas: Transhumanism," **Foreign Policy** 144
(2004): 42–43.

CHAPTER 12: SEARCH FOR EXTRATERRESTRIAL LIFE

363 **"We only have to look at ourselves":** Arthur C.
Clarke once said, "Either there is intelligent life
in the universe, or there is not. Either thought is
frightening."

363 **"If you live in a jungle":** Rebecca Boyle, "Why
These Scientists Fear Contact with Space Aliens,"
NBC News, February 8, 2017; www.nbcnews.co/
storyline/the-big-questions/why-these-scientists
-fear-contact-space-aliens-n717271.

365 **This is called SETI:** At present, there is no
universally accepted consensus concerning the
SETI Project. Some believe that the galaxy may
be teeming with intelligent life. Others believe
that perhaps we are alone in the universe. With
only one data point to analyze (our planet), there

are very few rigorous guidelines to direct our analysis, other than the Drake equation.

For another opinion, see N. Bostrom, "Where Are They: Why I Hope the Search for Extraterrestrial Intelligence Finds Nothing," **MIT Technology Review Magazine**, May/June 1998, 72–77.

392 **But all this still leaves one persistent, nagging question:** E. Jones, "Where Is Everybody? An Account of Fermi's Question," **Los Alamos Technical Report** LA 10311-MS, 1985. See also S. Webb, **If the Universe Is Teeming with Aliens . . . Where Is Everybody?** (New York: Copernicus Books, 2002).

393 **"Some of these pre-utopian worlds":** Stapledon, **Star Maker** (New York: Dover, 2008), p. 118.

395 **Another possibility is that they want to steal the heat:** There are many other possibilities that cannot be easily dismissed. One is that perhaps we are alone in the universe. The argument here is that we are finding more and more Goldilocks zones, meaning that it becomes increasingly difficult to find planets that can fit within all these new Goldilocks zones. For example, there is a Goldilocks zone for the Milky Way galaxy. If a planet is too close to the center of the galaxy, there is too much radiation for life to exist. If it is too far from the center, then there are not enough heavy elements to create the molecules of life. The argument is that there might be so many Goldilocks zones, many of them not

even discovered yet, that there might be only one planet in the universe with intelligent life. Each time there is another Goldilocks zone, it vastly decreases the probability of life. With so many of these zones, the collective probability of intelligent life is nearly zero.

Also, it is sometimes said that extraterrestrial life may be based on entirely new laws of chemistry and physics that are far beyond anything we can create in the laboratory. Hence, our understanding of nature is simply too narrow and simplistic to explain life in outer space. This may be true. And it is certainly true that entirely new surprises will be found once we explore the universe. However, it does not further the debate to simply state that alien chemistry and physics might exist. Science is based on theories that are testable, reproducible, and falsifiable, so simply postulating the existence of unknown laws of chemistry and physics does not help.

CHAPTER 13: ADVANCED CIVILIZATIONS

399 **The tabloid headlines blared:** See David Freeman, "Are Space Aliens Behind the 'Most Mysterious Star in the Universe'?" **Huffington Post,** August 25, 2016; www.huffingtonpost .com/entry/are-space-aliens-behind-the-most -mysterious-star-in-the-universe_us_57bb5537 e4b00d9c3a1942f1. See also Sarah Kaplan, "The Weirdest Star in the Sky Is Acting Up

Again," **Washington Post,** May 24, 2017; www
.washingtonpost.com/news/speaking-of-science/
wp/2017/05/24/the-weirdest-star-in-the-sky-is
-acting-up-again/?utm_term=.5301cac2152a.

400 **"We'd never seen anything like this star":** Ross
Anderson, "The Most Mysterious Star in Our
Galaxy," **The Atlantic,** October 13, 2015; www
.theatlantic.com/science/archive/2015/10/the
-most-interesting-star-in-our-galaxy/41023.

402 **This classification of advanced civilizations
was first proposed:** N. Kardashev, "Transmission
of Information by Extraterrestrial Civilizations,"
Soviet Astronomy 8, 1964: 217.

418 **"The premise is that any highly advanced
civilization":** Chris Impey, **Beyond: Our Future
in Space** (New York: W. W. Norton, 2016), pp.
255–56.

418 **"Logic tells me that it is reasonable to look
for godlike signs":** David Grinspoon, **Lonely
Planets** (New York: HarperCollins, 2003),
p. 333.

440 **The LHC has made many headlines:** It is
sometimes said that creating giant accelerators,
like the LHC and beyond, will create a black
hole that might destroy the entire planet. This is
impossible for several reasons:

First, the LHC cannot create enough energy
to create a black hole, which requires energies
comparable to those of a giant star. The energy
of the LHC is that of subatomic particles,
much too small to open a hole in space-time.
Second, Mother Nature bombards the Earth

with subatomic particles more powerful than those created by the LHC, and the Earth is still here. So subatomic particles with energies greater than the LHC are harmless. And lastly, string theory predicts that there might be mini black holes that one day might be found with our accelerators, but these mini black holes are subatomic particles, not stars, and hence pose no danger at all.

446 **Currently the only one capable of doing this:** If we naïvely try to join the quantum theory with general relativity, we find mathematical inconsistencies that have stumped physicists for almost a century. For example, if we calculate the scattering of two gravitons (particles of gravity), we find that the resulting answer is infinite, which is meaningless. Hence, the fundamental problem facing theoretical physics is to unify gravity with the quantum theory in a way that gives finite answers.

At present, the only way known to eliminate these troublesome infinities is to use superstring theory. This theory has a powerful set of symmetries in which the infinities cancel each other out. This is because in string theory every particle has a partner, called a "sparticle." The infinities coming from ordinary particles cancel precisely against the infinities coming from the sparticles, and hence the entire theory is finite. String theory is the only theory in physics that selects out its own dimensionality. This is because the theory is symmetric under supersymmetry. In general, all particles of the universe come in two types, bosons (which have integer spins) and fermions (which have half-integer spins). As the number of dimensions of space-time increases, the number of these fermions and bosons also increases. In general, the number of fermions rises much faster than the number of bosons. The two curves cross, however, at ten dimensions (for strings) and eleven dimensions (for membranes, like spheres and bubbles). Hence the only consistent supersymmetric theory is found in ten and eleven dimensions.

If we set the dimension of space-time at ten, then we have a consistent theory of strings. However, there are five different types of string theories in ten dimensions. For a physicist, searching for the ultimate theory of space and time, it is hard to believe that there should be five different self-consistent string theories. Ultimately, we want just one. (One of the guiding questions asked by Einstein was, Did God have a choice in making the universe? That is, Is the universe unique?)

Later, it was shown by Edward Witten that these five string theories can be unified into a single, unique theory if we add one more dimension, making it eleven-dimensional. This theory was called M-theory, and it contains membranes as well as strings. If we start with a membrane in eleven dimensions, and then we reduce one of these eleven dimensions (by flattening it, or slicing it), then we find that there are five ways in which a membrane can be reduced to a string, giving us the five known string theories. (For example, if we flatten a beach ball, leaving only the equator, then we have reduced an eleven-dimensional membrane down to a ten-dimensional string.) Unfortunately, the fundamental theory behind M-theory is totally unknown, even today. All we know is that M-theory reduces down to each of the five different string theories if we reduce eleven dimensions down to ten, and that, in the low-energy limit, M-theory reduces down to eleven-dimensional supergravity theory.

467 **If you then kill your grandfather before you are born:** Time travel poses yet another theoretical problem. If a photon, a particle of light, enters the wormhole and goes back in time a few years, then years later it can reach the present and reenter the wormhole once again. In fact it can reenter the wormhole an infinite number of times, and hence the time machine will explode. This is one of Stephen Hawking's objections to time machines. However, there is a way to escape this problem. In the many-worlds theory of quantum mechanics, the universe constantly splits in half into parallel universes. Therefore, if time is constantly splitting, it means that the photon only goes back in time once. If it reenters the wormhole, it is simply entering a different parallel universe, and hence it only makes a single pass through the wormhole. In

this way, the problem with infinities is solved. In fact, if we adopt the idea that the universe is constantly splitting into parallel realities, then all the paradoxes of time travel are solved. If you kill your grandfather before you are born, you have simply killed a grandfather in a parallel universe who resembles your grandfather. Your own grandfather in your universe was not killed at all.

CHAPTER 14: LEAVING THE UNIVERSE

481 **In the fifth epoch, even black holes:** Even black holes must eventually die. According to the uncertainty principle, everything is uncertain, even a black hole. A black hole is supposed to absorb 100 percent of all matter that falls into it, but this violates the uncertainty principle. Hence, there is actually a faint radiation that escapes from a black hole, called Hawking radiation. Hawking proved that it was actually a black body radiation (similar to the radiation emitted by a molten piece of metal) and therefore has a temperature associated with it. You can calculate that, over aeons, a black hole (which is actually gray) will emit enough radiation that it will no longer be stable. Then the black hole disappears in an explosion. So even black holes will eventually die.

If we assume that the Big Freeze takes place at some future time, we have to confront the fact that atomic matter as we know it might disintegrate trillions upon trillions of years from now. At present, the Standard Model of subatomic particles says that the proton should be stable. But if we generalize the model to try to unify the various atomic forces, we find that the proton may eventually decay into a positron and a neutrino. If this is true, then it means that matter (as we know it) is ultimately unstable and will decay into a mist of positrons, neutrinos, electrons, et cetera. Life probably cannot exist under these harsh conditions. According to the second law of thermodynamics, you can only extract usable work if there is difference in temperature. In the Big Freeze, however, temperatures drop to near absolute zero, so there is no more difference in temperature from which we can extract usable work. In other words, everything comes to a halt, even all possible life-forms.

481 **What is causing this sudden change in our understanding:** Dark energy is one of the greatest mysteries in all of physics. Einstein's equations have two terms that are generally covariant. The first is the **contracted curvature tensor,** which measures the distortions in space-time caused by stars, dust, planets, et cetera. The second term is the **volume of space-time.** So even the vacuum has energy associated with it. The more the universe expands, the more vacuum there is and hence the more dark energy available to create even more expansion. In other words, the rate of expansion of the vacuum is proportional to the amount of

vacuum there is. This, by definition, creates an exponential expansion of the universe, called de Sitter expansion (after the physicist who first identified it).

This de Sitter expansion may have given rise to the original inflation that initiated the Big Bang. But it is also causing the universe to expand exponentially once again. Unfortunately, physicists are at a loss to explain any of this from first principles. String theory comes closest to explaining dark energy, but the problem is that it cannot predict the precise amount of dark energy in the universe. String theory says that, depending on how you curl up ten-dimensional hyperspace, one can obtain different values for dark energy, but it does not predict precisely how much dark energy there is.

486 **The final possibility is to create a wormhole:** Assuming that wormholes are possible, there is still another hurdle to negotiate. One must be sure that matter is stable on the other side of the wormhole. For example, the reason why our universe is possible is because the proton is stable, or at least so stable that our universe has not collapsed down to a lower state in the 13.8 billion years of its existence. It is possible that the other universes in the multiverse may have a ground state in which, for example, the proton can decay to an even lower-mass particle, such as a positron. In this case, all the familiar chemical elements of the periodic table will decay, and this universe will consist of a mist of elections and

neutrinos, unsuitable for stable atomic matter. So one must take care to enter a parallel universe in which matter is similar to ours and is stable.

488 **All these speculations at first seem preposterous:** A. Guth, "Eternal Inflation and Its Implications," **Journal of Physics A** 40, no. 25 (2007): 6811.

488 **Also, when we look in one direction:** Inflationary theory answers several puzzling aspects of the Big Bang. First, our universe seems to be extremely flat, much flatter than usually proposed in the standard Big Bang theory. This can be explained by postulating that our universe had an expansion much faster than previously thought. A tiny portion of the original universe then inflated enormously and was flattened out in the process. Second, the theory explains why the universe is much more uniform than it should be. By looking in all directions in space, we see that the universe is quite uniform. But (because the speed of light is the ultimate velocity) there was not enough time for the original universe to mix thoroughly. This can be explained by assuming that a tiny piece of the original Big Bang was in fact uniform, but that uniform piece was inflated to give the uniform universe of today.

Beyond these two achievements, the inflationary universe theory so far agrees with all the data coming in from the cosmic microwave background. This does not mean that the theory is correct, only that it agrees with all the cosmological data so far. Time will tell if the

theory is correct. One glaring problem with
inflation is that no one knows what caused
it. The theory works fine after the instant of
inflation but says absolutely nothing about what
caused the original universe to inflate.

SUGGESTED READING

Arny, Thomas, and Stephen Schneider. **Explorations: An Introduction to Astronomy.** New York: McGraw-Hill, 2016.

Asimov, Isaac. **Foundation.** New York: Random House, 2004.

Barrat, James. **Our Final Invention: Artificial Intelligence and the End of the Human Era.** New York: Thomas Dunn Books, 2013.

Benford, James, and Gregory Benford. **Starship Century:**

Toward the Grandest Horizon. Middletown, DE: Microwave Sciences, 2013.

Bostrom, Nick. **Superintelligence: Paths, Dangers, Strategies**. Oxford: Oxford University Press, 2014.

Brockman, John, ed. **What to Think About Machines That Think.** New York: Harper Perennial, 2015.

Clancy, Paul, Andre Brack, and Gerda Horneck. **Looking for Life, Searching the Solar System.** Cambridge: Cambridge University Press, 2005.

Comins, Neil, and William Kaufmann III. **Discovering the Universe.** New York: W. H. Freeman, 2008.

Davies, Paul. **The Eerie Silence.** New York: Houghton Mifflin Harcourt, 2010.

Freedman, Roger, Robert M. Geller, and William Kaufmann III. **Universe.** New York: W. H. Freeman, 2011.

Georges, Thomas M. **Digital Soul: Intelligent Machines and Human Values.** New York: Perseus Books, 2003.

Gilster, Paul. **Centauri Dreams.** New York: Springer Books, 2004.

Golub, Leon, and Jay Pasachoff. **The Nearest Star.** Cambridge: Harvard University Press, 2001.

Grinspoon, David. **Lonely Planets: The Natural Philosophy of Alien Life.** New York: HarperCollins, 2003.

Impey, Chris. **Beyond: Our Future in Space.** New York: W. W. Norton, 2016.

———. **The Living Cosmos: Our Search for Life in the Universe.** New York: Random House, 2007.

Kaku, Michio. **The Future of the Mind.** New York: Anchor Books, 2014.

————. **The Physics of the Future.** New York: Anchor Books, 2011.

————. **Visions: How Science Will Revolutionize the 21st Century.** New York: Anchor Books, 1999.

Kasting, James. **How to Find a Habitable Planet.** Princeton: Princeton University Press, 2010.

Lemonick, Michael D. **Mirror Earth: The Search for Our Planet's Twin.** New York: Walker and Co., 2012.

————. **Other Worlds: The Search for Life in the Universe.** New York: Simon and Schuster, 1998.

Lewis, John S. **Asteroid Mining 101: Wealth for the New Space Economy.** Mountain View, CA: Deep Space Industries, 2014.

Neufeld, Michael. **Von Braun: Dreamer of Space, Engineer of War.** New York: Vintage Books, 2008.

O'Connell, Mark. **To Be a Machine: Adventures Among Cyborgs, Utopians, Hackers, and the Futurists Solving the Modest Problem of Death.** New York: Doubleday Books, 2016.

Odenwald, Sten. **Interstellar Travel: An Astronomer's Guide.** New York: The Astronomy Cafe, 2015.

Petranek, Stephen L. **How We'll Live on Mars.** New York: Simon and Schuster, 2015.

Sasselov, Dimitar. **The Life of Super-Earths.** New York: Basic Books, 2012.

Scharf, Caleb, **The Copernicus Complex: Our Cosmic Significance in a Universe of Planets and Probabilities.** New York: Scientific American/Farrar, Straus and Giroux, 2015.

Seeds, Michael, and Dana Backman. **Foundations of Astronomy.** Boston: Books/Cole, 2013.

Shostak, Seth. **Confessions of an Alien Hunter.** New York: Kindle eBooks, 2009.

Stapledon, Olaf. **Star Maker.** Mineola, NY: Dover Publications, 2008.

Summers, Michael, and James Trefil. **Exoplanets: Diamond Worlds, Super Earths, Pulsar Planets, and the New Search for Life Beyond Our Solar System.** Washington, D.C.: Smithsonian Books, 2017.

Thorne, Kip. **The Science of "Interstellar."** New York: W. W. Norton, 2014.

Vance, Ashlee, and Fred Sanders. **Elon Musk: Tesla, SpaceX, and the Quest for a Fantastic Future.** New York: HarperCollins, 2015.

Wachhorst, Wyn. **The Dream of Spaceflight.** New York: Perseus Books, 2000.

Wohlforth, Charles, and Amanda R. Hendrix. **Beyond Earth: Our Path to a New Home in the Planets.** New York: Pantheon Books, 2017.

Woodward, James F. **Making Starships and Stargates: The Science of Interstellar Transport and Absurdly Benign Wormholes.** New York: Springer, 2012.

Zubrin, Robert. **The Case for Mars.** New York: Free Press, 2011.

ILLUSTRATION CREDITS

INDEX

Page numbers in **bold** refer to illustrations. Page numbers beginning with 501 refer to end notes.